HOW TO DO
THINGS

HOW TO DO
THINGS

A TIMELESS GUIDE TO A
SIMPLER
LIFE

EDITED BY WILLIAM CAMPBELL
FOREWORD BY BRIAN BARTH

CHRONICLE BOOKS
SAN FRANCISCO

Compilation copyright © 2019 by Maria Ribas Literary, LLC.

All rights reserved. No part of this book may be reproduced in any form without written permission from the publisher.

Library of Congress Cataloging-in-Publication Data

Names: Campbell, William (Journalist), editor. | Atkinson, Wilmer, 1840–1920, editor.
Title: How to do things / edited by William Campbell ; foreword by Brian Barth.
Other titles: Farm journal.
Description: San Francisco : Chronicle Books, [2019] | Lightly edited version of a work originally published in Philadelphia by The Farm journal in 1919 and edited by Wilmer Atkinson. | Based on articles formerly published in the Farm journal.
Identifiers: LCCN 2018019520 | ISBN 9781452171678 (hardcover : alk. paper)

Subjects: LCSH: Agriculture—Handbooks, manuals, etc. | Home economics—Handbooks, manuals, etc. | Formulas, recipes, etc.—Handbooks, manuals, etc.
Classification: LCC S501.H815 2019 | DDC 635—dc23 LC record available at https://lccn.loc.gov/2018019520

Manufactured in China.

Design by Sara Schneider

10 9 8 7 6 5 4 3 2 1

Chronicle Books LLC
680 Second Street
San Francisco, CA 94107
www.chroniclebooks.com

For Brian Braun,
a man who knows how to do things.

A FARM JOURNAL FARM

Our Folks—by which is meant the four million subscribers and readers of "The Farm Journal"—live in homes such as the one pictured above. They are the cream of the agricultural people of America. They have solid, substantial homes, big barns with tight roofs, money in bank, fertile land, the best stock, the biggest apples, the richest milk. It is for their benefit that "The Farm Journal" is published, and to them this volume "How To Do Things" is offered.

AND HERE WE HAVE—

PETER TUMBLEDOWN'S FARM

No one can read "The Farm Journal" and be a Peter Tumbledown too. Many have tried, but they have to give up one or the other.

CONTENTS

FOREWORD:

PRACTICAL,
not fancy, LIVING

I am a writer and a journalist, which means I spend most of my waking hours staring at a screen. I tap away on my phone following the daily news cycle on Twitter. I fire off dozens of emails each day to my editors and sources. I comb the webscape to the point of exhaustion in search of any nuggets of information that might prove useful for whatever story I'm working on. I occasionally write early drafts by hand, but it's hard to resist the allure of Microsoft Word—cutting and pasting, the backspace key, hyperlinks, and Track Changes aren't an option when working with pen and paper.

Like most folks in the modern world, I am deeply, absurdly, worrisomely entrenched in the virtual abyss.

Which is why on a daily basis I must—*must*—command my hands to stop typing and maneuver the plastic, hairless mouse over to the sleep key on my monitor. Then I swivel in my office chair toward the door and bolt for that alternate reality where the earthworms still dwell, where things smell of woodsmoke, where milk comes from animals in a hay-scented barn out back, where time is measured in lambing seasons and frost dates, not news cycles and deadlines, and where mice are creatures of flesh and blood.

This is the world of *How to Do Things,* and it is alive and well.

Most Americans left this world behind over the course of the last century, but it did not cease to exist. Those who did not leave it are regarded, at best, as charmingly eccentric; *backward, naïve,* and *boring* are other adjectives often hurled. I wonder if those who look snidely down their noses at people who think chopping wood is a great way to spend a Saturday afternoon know what they're missing. As enthralling as Facebook and the latest Netflix series may be, I'd argue that this alternate reality is the more riveting of the two. It is certainly the more real one.

I travel no farther than my backyard to get there, where I have, over the years, chopped my own firewood, raised goats and pigs, cut fence posts, dug a pond, built a henhouse, achieved a measure of self-sufficiency in vegetables, and had far too many adventures to list—though let's just say most of them left me dirty, wet, scratched up with briars, smelling of earth and sweat, and happier than I've ever felt while sitting in front of a computer screen. A half hour daily dose of this world is the minimum I need to stay sane.

The real challenge in "getting there" comes from the mental exercise required. That's where I think *How to Do Things* can help. Getting back to the place of mending fences and milking goats means peeling yourself away from a world that doesn't want to let go. Consider this book a rope to help you escape. It is a window that shows you where to go, a guidepost on the path back to the land.

In 1919, the year *How to Do Things* was originally published, roughly 50 percent of Americans lived in a rural area. Today, only 20 percent do, and less than 2 percent can rightly call themselves a farmer—a century ago more than 6 million Americans proudly wore that hat. In 1919, virtually every rural household, whether they sold crops for a living or not, kept at least a few livestock, tended a garden, put up vegetables for winter, and maintained a woodlot for firewood. What was then a necessity feels like a luxury today.

Rural Americans grew up knowing how to do these things, but they were always on the hunt for better ways to do these things. Many looked to *Farm Journal*, the magazine whose editors published the book you now hold in your hands, for advice. The advice they got was well seasoned, and it came with heaping helpings of wit and wisdom, which kept folks coming back for more.

Farm Journal was where you learned of the new and improved seed varieties and alternative plowing techniques recently developed at the nation's land-grant universities. The content was by no means limited to farming. Cooking, entertaining, and cleaning advice cropped up on the magazine's pages, as did politics and poetry; even the occasional cartoon could be found. *Farm Journal* was the definitive rural lifestyle magazine of the early twentieth century, its unpretentious attitude summed up in the tagline beneath the title: "Practical Not Fancy Farming."

Much of *Farm Journal*'s magic, it seems, flowed from its idiosyncratic founder, publisher, and editor in chief for forty-plus years, Wilmer Atkinson. A son of Quaker farmers from Bucks County, Pennsylvania, Atkinson channeled both the simplicity and eccentricity of rural culture into *Farm Journal* and into this tome, a collection of the magazine's most beloved bits of advice and quirkiest interludes. They are "full of snap and ginger," he writes, "and hit the nail squarely on the head."

Atkinson's editorial style alternates between playful child and wise elder. He invented an alter ego, Peter Tumbledown, a goofy soul whose ramshackle farm is what you'll end up with should you fail to heed *Farm Journal* advice, as a gentle reminder to always close the barn door and never leave your tools in the rain. Most of all, Atkinson channeled love and tenderness for rural communities into every page, always referring to his readers as "Our Folks."

In this reprinting of *How to Do Things*, which has been only lightly edited and curated from the original, you'll learn how to make a homemade hog scratcher ("a device that will take the lice off [your] hogs as they are sound asleep") and protect the orchard from frost (in which you'll discover the importance of an "orchard alarm thermometer"). Recipes range from brown sugar pickles (one pound of sugar per dozen large pickles, plus "a few cloves and a little stick of cinnamon") to popcorn pudding (which will "please the children"). Some entries are irresistibly obscure (throwing a "cat party"), while others are intriguingly abstract (see "Put Your Wits to Work").

The book is full of anachronisms, many of them delightful, some a crude reminder of the rigid and inequitable social mores of the day—turns of phrase like *housewifely arts* are not uncommon. Atkinson was a product of his time, but he was also a bit ahead of his time: he served as president of the Pennsylvania Men's League for Woman Suffrage, which operated out of the *Farm Journal* office in Philadelphia. "I was an ardent friend of the cause . . . all my life," he wrote in his autobiography.

Atkinson passed away the year after *How to Do Things* was published and just a few months before the Nineteenth Amendment granted women the right to vote. At that time, *Farm Journal* was one of the most widely read magazines in circulation, with more than a million subscribers. To understand the breadth of *Farm Journal*'s reach, consider that prominent publications like *GQ* and the *New Yorker* boast a similar number today—when the population of the country is three times greater.

To be sure, neither *GQ* nor the *New Yorker*, nor any title you're likely to find on the top rack of the newsstand, will inspire you to purchase a herd of goats and take up cheese making. Today we don't need to know how to do these things. Then again, in a different way, maybe we do.

—Brian Barth

INTRODUCTION:
INTERESTING *and*
IMPORTANT MATTERS

In introducing this volume to the millions of readers of *Farm Journal*—the great family of rural Americans whom we often call with respect and affection "Our Folks"—only a very few words are necessary.

At intervals for the last forty years we have had friends of *Farm Journal* come to the editors and say, "Why don't you make a book of the best things out of the paper? It is a shame," they say, "to have this wealth of information lost, or to force readers to make scrapbooks to preserve special articles that interest them."

The editors have always agreed that such a book would be a good thing. From their daily contact with subscribers all over the country (via the countless letters and inquiries they receive and answer), they know very closely what folks are interested in and want to know about. On this contact they are accustomed to base the choice of what goes into *Farm Journal*, and they have always believed that a book containing these important things would be appreciated by many.

It was more a question of finding the time and energy to do the job. Other matters pressed, the family of Our Folks kept getting bigger and bigger, and demanding more and more information and advice, and so on and so on. However, the time came at last, and here finally is the long-desired book, full to overflowing with interesting and important matters.

On the editorial page of *Farm Journal* you may read under the title the words, "Unlike Any Other Paper." This states a truth that is obvious to those who form the habit of reading our journal—and which applies in equal measure to this book.

For one thing, it says a thing and stops after it has said it. For another, it is cheerful and hopeful and wants everybody to have a good time. It is young in spirit if not in years. For another, it is full of snap and ginger and hits the nail squarely on the head. Hence, its columns are perfectly clean and pure; it does not have to be hidden from the children, nor carried out of the house with tongs. For another, it prints no long-winded, tiresome essays and knows what to leave out as well as what to put in.

It is at home in every state, in all latitudes and longitudes, and welcome from the rising to the setting of the sun. It treats every branch of farming and living: gardening, stockbreeding, dairying, poultry raising, bee culture, sheep and swine husbandry, fruit growing, and trucking.

It thinks the humans of the farm are the best stock on it, yet it teaches the kindest, wisest care for every species of animal. Its teachings are practical and therefore profitable. Its purpose has always been, and now is, not how much profit will accrue, but how much good it can do. If our business were a mere matter of dollars and cents, we would quit today.

If we should sum up in a few words our feelings about *How To Do Things*, we would say something like this: a favorite description of *Farm Journal* is to say that it is "cream, not skim milk." The contents of this book have been selected as the very best of *Farm Journal* contents. We therefore present it to Our Folks, old and new, as the "cream of the cream."

—William Campbell and the editors

PART I

DOWN
on the
FARM

MAILING LISTS
that
MAKE MONEY

A good way to index customers

Almost every farm can use mailing lists to advantage. In buying, the lists help locate the cheapest and the most suitable article at once. In selling, they drum up trade, add new customers, and help obtain the highest market prices.

SELLING SEED CORN

A farmer who had made a hobby of growing sweet corn—cultivating it for years until he had developed a superior strain—found that the local stores were glad to handle his seed but offered a low price. He compiled a list of nineteen seedsmen operating in his territory, securing the names from farm paper and newspaper advertisements and from personal knowledge. Some of them sold seeds by mail, some through retail stores, and others were city wholesalers.

Using a typewriter, the farmer wrote a businesslike letter to all nineteen seedsmen, telling them what he had to offer and forwarding samples. His results were typical of mailing-list work. Eight businesses did not reply at all, and seven answered that they had adequate supplies arranged for. Four firms offered $1 a bushel more than the grower had been offered at home.

SELLING NUTS

Another farmer with a big crop of hickory nuts used a list of entirely different character. During a trip to town he borrowed several city directories and wrote down the names of professionals, manufacturers, and others that he believed had better-than-average buying power. These were classified in the directories and were easily copied. To every one of the several hundred families on his list, he mailed a printed catalog, with 1-cent postage, describing the superior sort of produce he had to offer and quoting prices in bulk lots. The prices were somewhat below what the city fruit stores were charging. He easily sold his entire crop in this way and had a fine beginning toward a parcel-post trade in other farm products.

SELLING WOOD

A third farmer obtained a list of pulpwood buyers and secured a price 50 cents a cord better than that which he was about to accept for his wood.

Purchasing agents of corporations send form letters to every business manufacturing articles they are in the market for, giving specifications and asking for samples and prices. Farmers who are making purchases of considerable size can follow the same plan. From the local farm newspapers a list of businesses can be readily complied. It is well to get prices from the local dealer, too.

Whether the articles to be bought are fence posts or farm implements, it pays to feel out the market thoroughly, and the mailing list is a cheap, effective way of going about it. The fellow who buys without comparison is often the disappointed one.

Mailing lists for most farm purposes can be compiled at home. There are businesses that make a specialty of furnishing lists, their charges running from about $2 for every thousand names with a guarantee of accuracy. If the list written to is a long one, it is oftentimes good business to use a printed form letter. With smaller lists, a typewriter will do. The typewriter lessens the labor in correspondence, and every farmer who does a great deal of writing should have one.

LEADING *the* BULL SAFELY

Where a bull is kept on the farm great care must be taken that he has no chance to do any one an injury. No chances should be taken. A rope attached to a ring in the nose serves as an extra hitching arrangement in the stall, but the bull should not be led by this alone. He can charge on the one leading him at will. Put an extra ring in the rope near his nose and have a stick with a snap in the end, and then the bull can be led anywhere in safety, the rope and the stick being taken together in the hand.

STOP CHICKEN EATING

Here is a cure for that old hog that eats up all the chickens. Use a piece of stiff leather wide enough to cover the hog's face within an inch or so of the snout, and secure it with a hog ring to the lower edge of the ears. An old bootleg will do.

ANOTHER HEN DISCOURAGER

Hiram Hogg: "At last my owner has solved the hen problem to my entire satisfaction by hinging the door to my sty so that it will always swing shut. When I leave my house to roam in the alfalfa I push it open with my snout and need not worry about any fussy old hen and a host of chirping chickens scratching in my nest. Nor will I again waken from my afternoon nap to find that same fussy old hen hovering her brood on my back."

HOMEMADE HOG SCRATCHER

Here is a device that will take the lice off the farmer's hogs as they are sound asleep. Drive a stake in the ground, wrap an old rope around the stake, and tack with shingle nails. Saturate the rope with equal parts of coal oil and lard once a week, or use one of the commercial coal-tar dips. Drive the stake near the hogs' sleeping quarters. This is so effectual that the hogs will stand in line waiting their turn to rub against this homemade hog scratcher.

SIMPLEST KIND
of
MILK STRAINER

Good butter making begins as far back as the milking, if not further. The process of milking must be clean if sweet butter is to be made. Fit a funnel, with strainer in the bottom, to the milk pail and milk into this. This will keep out much floating dust and will also assist in keeping the milk closed to odors while it has to remain in the stable.

RAISING GUINEA PIGS

The guinea pig is a native of Brazil and comes in three different colors—white, black, and fawn. Some of the white ones have red eyes.

Before starting in the business of raising guinea pigs, you should carefully consider several things:

If you have hay, apples, and similar feed on the home place, it is all right; if not, it may be a mistake to start in the guinea pigs business, as these feeds cost too much. Grain must be purchased, but that is a small expense compared with the other feed.

Then there must be a good place to keep the little animals. They won't thrive down in the cellar, nor out in the shed, nor up in the garret. They must have a place where a fire can be kept in cold weather.

They must be attended to as regularly as other farm animals. They must be watered once daily, fed two or three times, and have their hutches cleaned out every day.

When you get two hundred or three hundred guinea pigs, which would be necessary to have a steady income, you will find it work—not hard labor, but work you cannot shirk.

SCIENTIFIC HAND MILKING

ITS IMPORTANCE OFTEN OVERLOOKED

Too little attention is paid to the subject of scientific hand milking. A poor milker may easily do enough harm in a herd of cows in one year to equal in loss the amount of his wages. In other words, it would pay to hand him his year's salary in a lump sum and buy him off instead of allowing him to milk poorly ten or twelve cows each night and morning. Such a milker, if he is rough, cross, noisy, unclean, irregular, or imperfect in his milking, may quickly or gradually dry off the cows.

A SCHOOL WOULD BE GOOD

We know of one case in which a beginner, in two months, completely dried off the milk secretion in the cow upon which he was allowed to practice. In another case a new milker by his roughness and harshness so reduced the milk flow that the owner had to fire him in self-defense. It probably is a fact that in every herd where the milk is not weighed night and morning and close tally kept one or another of the milkers is doing indifferent or disastrous work. In Great Britain, girls who are taking up farm work are learning to milk by practicing on dummy cows until they become sufficiently expert to tackle the living animal safely. It would be well were our would-be hired hands put through such a course of training to make them proficient without spoiling or injuring a cow or two in the process.

THE MILKER OFTEN TO BLAME

Seeking the cause for many mysterious cases of intermittent garget—inflamed udders—experienced in some dairies, it must be suspected that the milker often is to blame. We think that incomplete milking is a possible cause, but one that is little suspected. The way a milker feels at milking time will in many instances determine the amount of milk he obtains. If he quickly extracts it all, it will be well for the cow and the employer. If he is in a hurry, indifferent, tired, or feeling sick and does not strip the cow clean, slight, unexplained garget may result. If such work continues, the cow will soon show a serious shrink in milk, prove profitless, or dry off entirely and have to be discarded.

It would be a good plan for every dairyman, especially in herds where slight cases of garget are prevalent, to have an expert strip the cows 10 minutes after the milkers have finished. By this means some very rich milk will be obtained for use on the farmer's table and at the same time a check will be kept on the work of each milker and some cases of garget possibly prevented. Knowing that the cows are going to be stripped, the milker will, if conscientious and anxious to please, milk just as well and completely as he knows how, and so all concerned will be benefited. If he is the other sort of a worker, he will be detected and discharged before he has done permanent damage.

FAST MILKING PAYS

The milker who can make the milk fairly boil in the pail and raise a lot of foam usually is getting the maximum flow of milk from each cow, while the slow milker, no matter how particular or faithful, often fails to get all that the cow would let down to the fast milking expert. A change of milkers may have a good or bad effect. In one experiment, two equally proficient milkers changed cows and at once there was an increase in milk yield from each lot of cows. A change of milkers, however, more commonly results in a decrease in milk production, and this sometimes is so noticeable that the accustomed milker has to resume his work with affected cows.

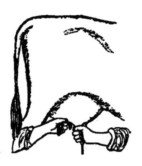

THE STABLE'S FLOOR
and
OTHER MATTERS

How is the stable's floor? Has it been pawed out in front and is there a space under the manger? A horse, the most ambitious animal when up, is the most helpless when down. Many a fine animal has lain down naturally enough, and in some manner, when trying to rise, has forced himself forward under the manger, or trough, and been found dead or badly injured in the morning. All our folks know about this danger, but in the multiplicity of things to be looked after it is sometimes forgotten.

Let us consider the manger. To begin with, every horse ought to eat off the ground. If the bottom of the manger is on, or within 2 inches [5 cm] of the earth or floor, no horse can get under it, and where the knee touches the manger when the animal paws it will rarely be continued. Standing upon an earth floor is good for the feet and necessary for some horses, but such a floor must be leveled often.

WATCH THE COAT

A horse's coat is a good indication of his condition at this season of the year. If it "stares," or looks rough and unkempt regardless of the daily brushing, he is not fully nourished and needs a change of feed. A molasses addition to the ration, of say a ½ cup [160 g] or 1 cup [320 g] twice daily, or a small handful of oil meal gradually increased to 2 cups [480 ml] twice a day, or 8 cups [2 L] of potatoes or apples twice daily, will presently work wonders in his appearance and spirits. A warm bran mash once a week is also good.

BEWARE OF ICY SPOTS
Do not run risks on icy spots. It is better to carry half a peck of sand and make a gritty path for the horse rather than force him forward with dull shoes.

DENTIST WORK
I have seen two horses of late that plainly showed they had ground their teeth to a sharp edge and now were suffering with sore cheeks due to laceration. It is easy to smooth down the sharp and rough outer edges of the molars when they get in this condition. Gently draw out the tongue, hold it first on one side and look carefully at the teeth, then the cheek, and then next the tongue. Let the sun shine into the mouth so you can see plainly. Then hold the tongue on the other side and repeat the inspection. Next, after giving the horse a little rest, take a sharp file by the handle and rasp off the troublesome sharpness. If you doubt your ability to do this, employ a veterinarian.

SYSTEMS IN HARNESSING
Harnessing and unharnessing necessarily take up much time on every farm. But, on some farms time is wasted thus that might be saved. To stop this extra labor, let a carefully planned system be followed by all who handle the teams, in both harnessing and unharnessing, so that everyone shall know exactly where to find each strap and snap. Speaking of snaps, these useful little things need looking after once in a while, to see that they have not gotten out of order and so are ready to fall and perhaps make trouble and expense.

THE MOST MONEY
in
EARLY CHICKS

The best time for hatching future layers or breeding stock in most areas of the country is between the fifteenth of March and the fifteenth of May. This gives the youngsters a good start, and before the hot weather of July and August strikes them they will have matured sufficiently to be able to withstand the depression. Late-hatched chicks are likely to become stunted when hot weather comes. This causes a setback, and there will be few if any eggs before the latter part of February or March. Early-hatched young hens, or pullets—those that lay at or before eight months of age—are the ones that lay when eggs are scarce and prices high.

EARLY, BUT NOT TOO EARLY

The eggs from early chicks will be of good size. On the other hand, late-hatched young hens at times lay eggs that are so small they are practically unsalable. Early-hatched pullets are more steady layers, and their yield can be regularly counted on. However, it is not advisable to get pullets out before the sixteenth of March, because such birds will go into molt in the fall and thus pass over a valuable season without laying any eggs.

BETTER FOR MEAT

Aside from laying, there is a big value in the early chicks from a meat standpoint. Such birds will develop large carcasses and come into the broiler or the soft-roaster age in fine condition and just at a time when there is a strong demand and prices are high. From a breeding standpoint, we get better vigor, better laying, better fertility, and better size from hatches made between the sixteenth of March and the sixteenth of May than we do from later hatches; in many ways, too, they are better than hatches brought out during February and the early part of March.

GOOD FEED AND CARE ESSENTIAL

The value of early-hatched pullets and roosters, or cockerels, does not depend alone upon the time of the year they were hatched. Instead, good care and good feed go hand in hand with the date of incubation. In other words, they must be kept growing from the very beginning. Any setbacks will show themselves at once. There must be good, nourishing food, and it must be given so that the chicks do not stall at it. I believe in having a trough of dry bran or some commercial chick feed where the little ones may help themselves between meals. Part of their ration must be mash food and part cracked grains or commercial chick feed.

Equally important with the kind of feed is regularity in giving it. I never took kindly to the old advice of feeding every two hours. However, the chicks should be fed three times a day morning, noon, and night. There must be a regular hour for feeding and at that hour the food must be given. It is almost incredible how well the chicks, with their ravenous appetites, know when feeding time has come, and every moment's delay has a telling effect upon them.

With wise feeding comes exercise, which is induced by scattering the grain among the litter on the house floor. This exercise not only sharpens the appetite, but it puts the pullets in a good, vigorous condition. Early pullets, well hatched, well fed, and well cared for, will mature rapidly and go to laying at six to eight months of age, according to breed.

BIG-FOUR EGG FORMULA

The essential thing is to get the hens to eat enough of any combination to bring results. While the following proportions may not produce the highest results, they will certainly raise the general average of egg production wherever put to practice. There are many groups that might form a big-four combination. In this instance, the four most common feeds have been selected. The nutritive ratio, as given, will be found reliable for autumn and spring feeding. But for the winter period, 25 percent more cracked corn may be fed to advantage. For summer, wheat or first-grade screenings should be substituted for the corn. The quantities, as given, may be increased or decreased to accommodate the size of the flock.

> *Cracked corn: 49 pounds [22 kg]*
> *Wheat bran: 31 pounds [14 kg]*
> *Beef scrap: 13 pounds [6 kg]*
> *Mangel-wurzel beets: 29 pounds [13 kg]*

MIX FEED WITH EXERCISE

A scratching place should be prepared and supplied with clean straw to the depth of 1 foot [30.5 cm]. Scatter cracked corn in this area several times daily. This exercise will not only do good, but will add to the egg basket. Place the bran and beef scrap in an open hopper. Do not mix them. Reduce the beets or place in a clean spot for the hens to pick at. They may be hung up, just enough to force Biddy to jump for them.

OTHER CONDITIONS TO WATCH FOR

The most careful, systematic feeding in the world will not produce eggs satisfactorily unless other conditions are favorable. Your hens must have a warm, dry place for roosting and exercise. Especially at night they should be kept free from direct drafts of air. Cleanliness and absence of vermin are essential to success in egg production. This all means work, but it pays. Given the same attention as other things, the egg industry can be rendered as profitable as any other department of farming.

FEEDING STOCK
on
RANGE

The very best way of growing chickens, especially when intended for future breeding, is to give them free range and provide them with food just when they need it. The outdoor feed hopper for dry mash here given is so plain that it will not be difficult to understand making the hopper. A roof can be constructed over it to keep out the rain and hot sun. The hoppers are to be kept well filled with dry mash, and the youngsters, after taking vigorous exercise, are able to sharpen their appetites so that they eat greedily of the meal set for them. The hoppers can be made any size.

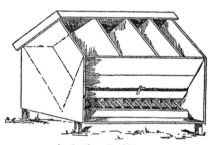

An Outdoor Feed Hopper

HOW TO TELL WHAT IS KILLING *the* CHICKENS

We may pretty well determine the character of the animal that visited our henhouse by the condition of the fowls as found. Should an opossum get into the coop he will kill but one or two on his visit. He eats the head and neck of the victim and doesn't seem to care for the rest of the carcass.

A mink is more deadly. He will slaughter a dozen or more birds in a night, biting them in the neck and sucking the blood. Both the mink and the opossum leave the carcasses in the coop or house where they found them.

Rats drag their prey into the holes or runways. Rats, however, very seldom attack a half-grown chicken or fowl. Their appetite is more for the youngsters, so the front of each coop should be closed with a wire-covered frame, which keeps out the rats and permits ventilation. Cats and foxes carry their victims away with them; the cat, like the rat, cares only for the baby chicks, seldom doing damage to birds that weigh more than 1 pound [455 g].

The skunk seems to select poultry for his diet only as a last resort. He prefers refuse meat or scrap. If any of the tatter is found, he will fill up with it and then retire to his den. The next night he will return, and in case the refuse meat or scrap is insufficient to satisfy his appetite, he will top off on poultry.

The weasel crawls on the roost, selects his victim, taps a vein, and sucks the blood. The weasel is a regular contortionist and is able to so contract his body that he can wedge through the smallest opening.

Where there are foxes, there will also abound opossums, minks, weasels, and skunks. None of the above pests should be allowed to harbor near the poultry yard. Piles of rail or stones, stone walls, and briar patches make a safe harbor for these vermin. A good active dog will do much to keep them away.

BIRDS OF PREY ARE DEADLY

Hawks are deadly enemies. They have the habit of perching nearby and surveying the territory and, having once laid plans, descend upon the young chicks. Crows, while not as a rule poultry enemies, will when very hungry attack chicks, especially in the spring when there are young crows in the nest to feed. Little chickens that run around the coops in which their hens are penned up are particularly apt to be victims of the crows. If the hen is running at large, she can usually protect them. The owl is a night bird, and frequently feasts upon large poultry roosting in trees.

FOXES

The fox has a cunning way of hiding in woodlands near fowls. He keeps a close watch and waits until some reckless hen wanders near him, then suddenly grabs his prey and carries it to his haunts. If unable to capture any poultry during the day, he will visit the hennery at night.

In areas where foxes are known to exist, the foundations of the poultry houses should begin 1 to 2 feet [30.5 to 61 cm] under the ground, as the sly old reynard has a knack of making a tunnel under the sill to gain entrance to the house. Care should also be taken that the houses are so closed at night that it is impossible for any of the night-prowling enemies named previously to enter. It is a wise poultry keeper who padlocks the door, for there are other poultry enemies besides those mentioned: the two-legged kind. There can be nothing more discouraging than to find, on opening up the pens in the morning, that some "midnight poultry raiser" has cleaned up your entire stock of chickens.

SIX BREEDS
of
MONEY-MAKING GEESE

The leading market breeds of geese are the Toulouse, Embden, wild or Canadian, African, brown Chinese, and white Chinese. The first two are probably the most popular. The Toulouse, the African, and the Canadian are gray in color; the Embden and white Chinese have white plumage; the other breed is brown.

An adult Toulouse gander weighs 26 pounds [12 kg]; the adult goose, 20 pounds [9 kg]; the young gander, 20 pounds [9 kg]; and the young goose 16 pounds [7 kg]. An adult Embden gander weighs 20 pounds [9 kg]; adult goose, 18 pounds [8 kg]; young gander, 18 pounds [8 kg]; young goose, 16 pounds [7 kg]. The weights of wild or Canadian adult gander, adult goose, young gander, and young goose, are 12 pounds [5.5 kg], 10 pounds [4.5 kg], 10 pounds [4.5 kg] and 8 pounds [4 kg], respectively. The adult birds of the African breed have the same weight as the Embden adult birds, but the African young gander and young goose weigh 16 [7 kg] and 14 pounds [6 kg], respectively. The Chinese geese have the same weights as the Canadian. The African and Chinese breeds show distinctive knobs or protuberances on their heads, while the heads of the other breeds are plain.

Toulouse geese have massive bodies of medium length, broad and very deep, almost touching the ground. They are good layers, producing twenty to thirty-five eggs a year; they are docile, grow rapidly, and are excellent birds for the market.

Embden geese are considered only fair layers, the egg yield varying greatly among individuals. As market birds, they mature early, grow rapidly, and on account of the white pin feathers are popular with dealers. They furnish more attractive carcasses than Toulouse.

Wild or Canadian geese are rather poor layers and are often difficult to breed successfully in captivity. They are crossed with other breeds to produce the so-called mongrel type, which is much prized for market purposes but is usually sterile.

African geese are good layers as well as good market geese. They grow rapidly and mature early.

There are two standard varieties of the Chinese goose (the brown and white), and both varieties mature early. As layers, they are prolific, and as market poultry are rapid growers. The Chinese geese are naturally shy and therefore rather difficult to handle.

Geese can be raised in small numbers where there is low, rough pastureland, with a natural supply of water. Grass makes up the bulk of their feed. While they can be kept without the

use of water to swim in, it will be all the better, especially during the breeding season, if a body of water is available, as it affords exercise that the geese could not obtain otherwise.

The period of incubation of goose eggs varies from twenty-eight to thirty days. The first eggs laid by the goose are usually set under chicken hens (as geese do not lay eggs and sit on eggs at the same time). The last eggs that the goose lays may be hatched either under chickens or under the goose, if she becomes broody.

The broody goose plucks off more or less down from her breast, with which to line the nest and cover the eggs whenever she leaves them. A goose can conveniently cover eleven eggs, but a chicken hen should not be given more than five.

Year-old geese are not mature enough for breeders. The females lay fewer eggs of smaller size, and usually more of them are infertile than is the case with females two or three years old.

The bill of the goose is provided with sharp, interlocking, serrated edges, designed to cut and divide vegetable tissues easily. The tongue, at the tip, is covered with hard, hair-like projections pointing toward the throat, which serve to convey the bits or grass and leaves into the throat quickly and surely.

A GOOD LIVING
and
10 PERCENT

We believe that every knowledgeable farmer who knows his business is entitled to "A Good Living and 10 percent," and when we are asked to define exactly what we mean, here is how we do it:

DEFINITIONS

1. "A Good Living": A farmer and his family should live comfortably and certainly not less well than a family in similar circumstances in a town or city.
2. "And 10 percent"—besides a good living: A farmer should earn at least 10 percent on the amount invested in the farm enterprise, which would be only 5 percent on the capital and only 5 percent actual profit besides.
3. "Farmer": The term *farmer* as used in this context means one who is engaged in any branch of agriculture as the chief means of livelihood.

MEASURING THE INCOME OF A FARMER

The problem is different from that of those engaged in most other enterprises, for in those cases people usually receive a definite salary for their labor, out of which they pay their living expenses. A farmer's income, however, is composed partly of "living expenses" taken from the farm and is only partially received in money.

1. Living expenses obtained from a farm include:
   (a) A house to live in,
   (b) usually fuel; and,
   (c) to a large extent, food.

2. The other income received by a farmer is the difference between receipts and expenses. Receipts include:
   (a) Income from sales.
   (b) Increases in inventory other than mere increases in price of land, improvements, and equipment.

Expenses include:
1. Labor, except that of farmer, spouse, and minor children.
2. Materials purchased, such as feed, seed, and fertilizer.
3. Repairs to fences, buildings, and machinery.
4. Miscellaneous, such as taxes, depreciation on buildings, stock, and equipment, rent (if tenant farmer), fire and farm insurance, interest on debts, losses by fire, diseases, hail, pests, etc.

In addition to these costs of production there must be an allowance for the living expenses not obtained from the farm. These include items of household and family upkeep, such as food not produced on the farm, clothing, recreation and travel, education, life insurance, benevolences, doctor, dentist, etc.

WHERE THE 10 PERCENT COMES IN
From the "farm income," or balance left from sales after deducting cost, a farmer should have a balance left of 10 percent on

investment after paying all the expenses listed above. But this reservation must be made: the responsibility for spending the income of 10 percent necessarily rests upon the farmer himself; and, if he has chosen to spend this in luxuries, the choice was his—our part of the task is to make sure that farmers generally get this "Good Living and 10 percent," so that the nation's future food supply shall be assured.

It must be remembered that the amount earned (10 percent) will not necessarily be represented by a cash balance of that amount at the end of the year. The cash balance will be more or less than 10 percent, according to whether the farmer has decreased or increased his investment, as shown by inventories taken at the beginning and at the close of the year.

ASSUMPTIONS

The following assumptions have been made:

1. That the farm is large enough to warrant the entire attention of at least one person.
2. That the farmer possesses average ability and gumption, together with stick-to-it-iveness to carry the undertaking through.
3. That the family is of average size.
4. That only such buildings and equipment as are adequate and necessary shall be on each farm—in other words, plain, not fancy farming.

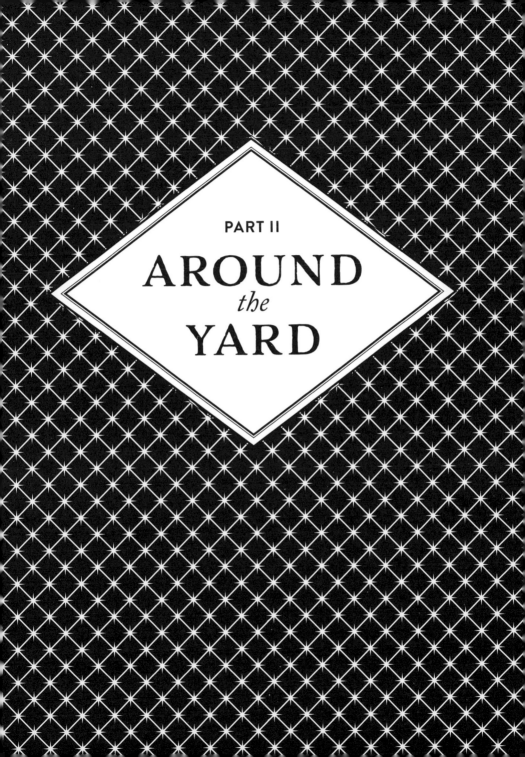

PART II

AROUND *the* YARD

PICKING APPLES

ELEVEN RULES THAT WILL ENSURE RESULTS

Some good apple-picking rules:

1. Pick from lower limbs first.
2. See that the ladder is pushed into the tree gently so as not to knock off or bruise the fruit.
3. Hang the basket so as to be able to pick with both hands.
4. Lay the apples in the basket; do not drop or throw them.
5. Pick no specked apples.
6. Pick no small, green apples.
7. Do not waste time picking a few little apples out of reach—let them go.
8. In emptying the basket, pour gently, as you would eggs.
9. Do not set one basket or crate on another so that the apples below are bruised.
10. Lift and set down gently all filled crates.
11. Use a spring wagon in hauling, avoid rough ground, and go slow except on smooth road.

THE RIGHT WAY
to
SAW A LIMB

To save time and labor in trimming trees, try sawing the underside of the limb about one-third of the way through or till the saw begins to pinch, and then saw on top about one-half inch in front of the under cut, and when you have sawed down almost to the under cut, the limb will break off and not peel down the side of the tree. One cut will do the same work as two cuts.

FRESH VEGETABLES ALL WINTER

A large proportion of the vegetables sold in Chicago in winter are preserved in outdoor pits. It is estimated that not less than ten thousand of these pits are inside the city limits of Chicago, and there are thousands of others in the county. Some of the pits are 30 feet [9 m] long and 10 to 12 feet [3 to 3.5 m] wide and hold more than 160 bushels of potatoes, roots, etc. Some are even larger and hold a carload of vegetables, keeping them out of reach of Jack Frost. That is practically the same system, but on a larger scale, as that used by our grandfathers to winter a small stock of garden vegetables.

HOW THE PITS ARE USED

The pits are sunk in well-drained soil to the depth of 15 to 18 inches [38 to 46 cm], and then the vegetables are carefully piled up to a cone, after which they are covered with a thick layer of soil 12 inches [30.5 cm] thick all over, then topped with a covering of straw or strawy manure.

When removing the vegetables to be hauled to market, one end of a pit is opened and the desired quantity removed. The stuff is sacked and then piled in a wagon box that has been thickly lined inside with gunnysacks or old comforters and blankets. Some of the vegetable sacks are kept in a cellar overnight and hauled to market the next morning and sold to consumers. Vegetables so preserved are crisp and fresh the winter through.

STRAWBERRY CULTURE

We wish we could prevail upon everybody who has even a patch of land to set out a bed of strawberries this spring, take good care of them through the season, and revel in this delicious fruit next summer. Surely, we cannot do our readers a better service than to persuade them to give immediate attention to this important matter. It is not a difficult thing for any person with even a small garden to grow strawberries in such abundance that every member of the family shall have enough for at least three weeks of the summer, for it is an easy fruit to grow and yields certainly and profusely in response to intelligent effort.

The strawberry bed should be started as early in the spring as soil can be got into mellow condition. In the latitude of Philadelphia, this occurs usually early in April—farther south, earlier; farther north, later. Let it be understood that it will not do to delay if the best results are to be attained. It will do to plant in early May, but not in early June nor late May. Select a piece of ground that is well drained, for the strawberry does not like wet feet (neither does it like dry ones). Old sod is not suitable, because it may harbor the white grub, which is very destructive to the roots of strawberry plants. A patch of ground that is likely to be as free as possible from weed seeds and is out of the way of chickens is, of course, best.

If you are a beginner and want to grow for market, ¼ acre [1,000 sq m] is enough to begin with; if only for family use, a bed 20 by 40 feet [6 by 12 m] is large enough to supply the family lavishly for nearly a month. Let no weeds grow until fruiting time, but pull, not hoe them out, for the ground should not be disturbed. Next summer be sure to have a good supply of cream on hand and send us an invitation to visit you. We will do the rest.

HARVESTING *the* ICE CROP

The larger the amount of ice packed in a structure, the better it will keep. It will not keep without drainage and ventilation. If there is a drain, it should include a trap to prevent air from entering the house. The icehouse should stand on sand or gravel or have ample artificial drainage put in to carry away the melted ice. The ice must be packed on a bed of sawdust or marsh hay 2 feet [61 cm] deep and be packed in a solid layer of ice cakes that are sawed with square angles and as large as can be handled conveniently, say 22 by 30 inches [56 by 76 cm]. Place these on edge, all one way and 12 inches [30.5 cm] from the sides of the building until solid, and if spaces occur between them, fill with sawdust.

After the upper surface has been leveled in the same manner, and the sides next to the wall filled with sawdust tamped hard, place the next layer of cakes. This is to be continued until the house is full, say to within 3 feet [91 cm] of the eaves. Over the top of the ice place 18 inches [46 cm] of sawdust with plenty of ventilation above.

Fig. 1

As mild weather approaches, the ice should be inspected two or three times a week and the side packing kept tamped hard to make sure that no ventilating tunnels occur in the sawdust, for these quickly waste a ton of ice. A stick may be needed to punch down sawdust into such spaces. As the sawdust is taken from the top to fill around the sides, more must be put on, keeping it 18 inches [46 cm] deep, but as ice is removed, see that the sawdust does not accumulate much deeper than that, as it will generate heat and the ice will not be ventilated and will waste rapidly.

Many persons in a rural community, upon learning that ice may be bought, will come and buy it, thus saving delivery. Of course, ice will wet a wagon if placed in it with no protection, but a sheet of galvanized iron, turned up at the sides and front by 1 inch [2.5 cm] and crafted to protect the underside as well, will keep the wagon dry.

FIGHTING POISON IVY

Poison ivy should not be allowed to go to seed, as it will contaminate the farm for years to come. It is frequently found along fencerows and roadsides, where it is too often neglected. It may be distinguished from Virginia creeper by the three leaflets per group as compared with the five leaflets of the Virginia creeper.

DESTROYING BY SULFURIC ACID

Concentrated sulfuric acid will kill poison ivy. Dose each plant with ½ teaspoon to each stem, making the application during the growing season, every 3 weeks. If a large area is covered by the plants, spraying with arsenic of soda (1 pound [455 g] of the poison to 20 gallons [76 L] of water) will kill all vegetation. One application, if the growth is young and tender, will do this.

DESTROYING BY FIRE

Here's another way to destroy poison ivy: a friend of ours puts straw along the stone fences and other areas infested with poison ivy, and then sets fire to the straw, repeating the operation at intervals until the plants give up trying to grow.

Poison ivy may be grubbed out by one who is immune to the poison. All parts of the plants should then be gathered into a pile and burned. The resulting smoke should not be inhaled nor allowed to get into the eyes.

CHASING *the* ROBINS

Robins bothering the strawberry patch in the garden? Well, here is one way to protect the produce. Kitty has been busy catching mice all winter, and it won't hurt her to have a brief change of occupation. Surely, she won't object if we put a nice collar around her neck, fasten her to a sliding tether attached to a taut overhead wire, and "turn her loose" to scare away the birds in that little berry patch. Not likely she will catch any, but she will give 'em a good scare.

WHEN BEES ARE PUT OUTDOORS

Beekeepers who wintered their bees outdoors in hives properly protected will have little to do with them in March, unless the weather is sufficiently warm at that point, as is often the case in southern states. Those who wintered their colonies in special repositories or cellars, however, will find the month of March to be an exceedingly busy one, for, generally speaking, March is the time to put the bees outdoors. Some progressive beekeepers, in places where the temperature during the winter gets below zero, wait until the pussy willows are in bloom before setting colonies out.

WRAPPING THE HIVES

In taking the colonies out of their winter quarters, it is best to give them some added protection in the form of telescope cases, after first wrapping the hives with old newspapers. Where cases are not at hand, the hives can be wrapped with old papers and the whole then wrapped with waterproof paper, tied in place, leaving the hive entrance open for air, etc.

SPRING DWINDLING

Reports from many beekeepers tell me that spring dwindling—when the majority of worker bees in a colony die off—is one of their greatest handicaps. They tell me that the colonies become so weak that they are not strong enough to secure much surplus

from the early flows, that it takes practically all spring and summer for the bees to recover from spring dwindling, and that they
become strong only in time for the late flow from buckwheat,
goldenrod, and aster.

I have in mind one beekeeper in New York State with
hundreds of colonies, whose whole surplus is practically from
buckwheat. He has told me on several occasions that the reason
he secures so little from the clover and basswood is because
of spring dwindling, which so reduces the number of bees in
each colony that it takes the colonies all summer to build up.
For colonies placed outdoors, our suggestions for added protection will in a large measure prevent the spring dwindling.

THE BUSY BEE
in MIDSUMMER

The bees are on the home stretch by midsummer, gathering much nectar from the clovers; and about the middle of July is the proper time to harvest the crop and keep it separate from the later fall flow of darker and not so richly flavored honey.

In the matter of giving the bees additional storage room and harvesting the clover honey, one cannot go entirely by the calendar, but must be governed by the conditions afield and the condition or the colony, as each colony will require individual treatment.

WATCH THE NECTAR SUPPLY

If the clovers show a tendency to stop secreting nectar, and here and there patches of it are beginning to die, then we should give no further supply or additional housing resources, but rather permit the bees to concentrate all their energies upon what sections they have, as it is folly to end the season with a large number of unfinished sections on hand.

If the season has been late, then the wise thing to do is to let the supers remain on even beyond the twentieth of the month, and be governed by the conditions prevailing afield.

UNCAPPING THE COMBS FOR EXTRACTING

In harvesting the honey from the extracting combs, a good sharp honey uncapping knife is a necessity. Two are better than one, as one can remain in a pan of hot water on the stove while the other is in use, and the knives changed with every two or three combs uncapped. Hot water not only cleans the knives of honey and wax, but keeping the knife hot helps it cut more easily. In any case the knives should be very sharp.

Holding the hot knife at just the right angle, it is a very easy matter to slice off the caps of the combs in the extracting frames and make them ready for extracting.

The honey and the wax that adhere to the knife can be scraped off into a can, and the honey separated from the wax by beating the honey until the wax melts and rises to the surface; when the mass is cooled the wax can be removed without trouble.

STORAGE

In harvesting the honeycomb, greater care must be exercised, and under no condition should the section combs be stored in a cold or damp place.

The cellar is the worst place imaginable for honeycomb, and an icebox is certain to make the honey sweat and run over the combs, which will spoil its appearance. The warm attic is an excellent place, but be careful to store it carefully so that no robber bees can gain access to it.

By all means use a bee escape board, allowing the bees to leave one way but preventing them from coming back in, in taking off the supers containing the sections.

TREE-PRUNING HINTS

Spring is a good time to prune trees, unless you prefer to wait until June (spring pruning induces wood growth and June pruning induces fruit growth). Of course, on young trees you should want only wood growth until they are good-sized and fully able to endure the strain of fruit bearing. Some growers do part of their pruning in March and part in June.

Don't prune mature trees too severely. A tree must have some branches upon which to produce its fruit; otherwise it will produce water sprouts instead of fruit.

Don't prune off a single branch unless you know just why you are removing it and why you are removing that particular branch in preference to some other.

Don't neglect to paint all large wounds. Painting will improve the appearance, prevent decay, prevent evaporation of the tree's supply of moisture, and facilitate healing.

Above all, *don't* allow anyone to prune your trees if his chief recommendation is his ability to handle an axe and a saw.

Don't prune your trees because someone else thinks they need pruning. He may not know any more about them than you do.

Don't prune your trees unless you can tell the difference between a dead and a living branch, between a bearing and a nonbearing branch, between a fruit spur and a water sprout, and between a fruit bud and a leaf bud.

Don't prune off the large limbs when equally good results can be had by removing a few of the smaller limbs. The large ones form the framework of the tree and are needed to support the bearing branches.

It is sometimes stated that the fruit growers of the Pacific Slope, who produce some of the finest fruit in the world, prune away "nearly half of the tops of their trees" every year. They do nothing of the kind. They remove one-quarter to two-thirds of the annual growth of the previous season. But they give their trees culture that causes the trees to make a terminal growth of 2 to 3 feet [61 to 91 cm] and often 4 feet [1.2 m]. The average eastern farmer gives his trees only enough care to permit the growth of 4 to 5 inches [10 to 12 cm] of terminal growth; and so his treetops do not need the same treatment that a larger growth would require.

Trim fruit trees a little every year, rather than a lot in any one year. Peach trees require more pruning than most trees; at least one-half of the length of the new growth should be removed each season. Cherry trees require the least pruning; merely cut out dead, broken, or "crossed" limbs. Other trees need a judicious thinning out and, sometimes, cutting back. Avoid cutting so as to leave "stubs"; make neat cuts close to union.

The harder you prune, the more suckers you will have; don't overdo a good thing.

PROTECTING *the* ORCHARD *from* FROST DAMAGE

The heavy demand for fruit at profitable prices, and the fact that Europe must depend upon our orchards for her supply, due to the depletion of her famous groves, should awaken fruit growers to the importance of making arrangements to insure their fruit growth against frosts in the early spring.

The most practical and economical method yet devised for protection of large areas is the direct addition of heat by means of numerous small fires properly distributed over the area to be protected.

WHEN TO START THE FIRES

An orchard alarm thermometer obviates the inconvenience of remaining on watch. This instrument is so delicate that an alarm bell sounds in the owner's bedroom when the orchard temperature approaches the danger point. The instrument is placed at what is considered the coldest or lowest part of the orchard, and is set so as to ring when the temperature is a few degrees above the danger zone. This allows time in which to light fires and get the heat up to the protection point before the frost becomes dangerous.

When a severe drop in the temperature is accompanied by a heavy wind, it is almost impossible to accomplish anything in the way of heating the orchard sufficiently to ward off the cold. But, fortunately, ordinary frosts occur during still weather.

FUEL

Several kinds of fuel are adapted for use as orchard heat. Wood is very good where it is plentiful enough. Coal, petroleum coke, and fuel oil all have given satisfaction, depending on the worth of the crop. J. G. Gore, a successful fruit grower in Oregon, saved his crop, which sold for $1,000 per acre [4,000 sq m], for several years by the use of old rails and stumps for fuel. A little dash of kerosene or crude oil, and the application of a torch, starts the fire. An iron rod about 4 feet [1.2 m] long, wound at one end with cotton waste or rags saturated with oil, makes a good torch.

Anywhere from twenty-five to fifty fires per acre [4,000 sq m] are needed, according to the size of the trees and the degree of frost. Fires need not be built large, as these tend to create currents that draw in cold air from outside and thus defeat the purpose. Carefully prepare the piles of wood beforehand, every other one being set off first, and those in between held in reserve for further use if the first are insufficient. In this manner, it is possible to hold the temperature of the orchard from six to ten degrees above the temperature outside. When the frost is severe, a dense smudge should be made about sunrise, which serves to retard the process of thawing until the sun has warmed the atmosphere.

OIL NOW THE FAVORITE

Oil has been more satisfactory with commercial orchardists who have a large area to fire. Oil is more quickly handled, either crude or distilled being acceptable. The crude is cheaper but is less satisfactory because it contains a higher percentage of water, which tends to extinguish the flame and cause the firepots to boil over. It is more difficult to handle in cold weather and in burning gives off large quantities of soot. Distillate, having no water content, ignites freely, burns readily, and creates little residue.

Firepots for burning oil may be had of any orchard supply house at reasonable terms, in quantities. Fifty to one hundred are used per acre [4,000 sq m]. Oil should be on hand so that it may be used quickly, as the success of this system depends on being ready for business when there is a frost indication.

PROTECT YOUR TREES *from* MICE *and* RABBITS

November is the time to protect orchard trees against injury from mice and rabbits. To prevent field mice, soil mounding is recommended. The earth should be mounded to a height of 6 to 8 inches [15 to 20 cm] and 12 to 16 inches [30.5 to 40.5 cm] in diameter around the base of each tree and well tamped down. All grass and litter should be cleaned away from the trees.

Cylinders of ¼-inch [6 mm] mesh galvanized iron wire are a good protection against either mice or rabbits. Pieces of such wire 12 by 24 inches [30.5 by 61 cm] make cylinders of a convenient size for small trees. The cylinder, after its edges have been fastened together, should be slightly imbedded in the ground to secure it.

Wrong—sinfully wrong!

The right way

Many protective washes have been suggested from time to time, but most of them have not proved satisfactory. Extensive experiments have shown the ordinary lime-sulfur mixture to be quite satisfactory. It is used at the ordinary strength, as for scale insects. The trunks of the trees should be sprayed or painted close to the ground and to a height of 2 feet [61 cm] above it.

A wash recommended by Ohio fruit growers is made of one peck of fresh limestone slaked with old soapsuds and the mixture thinned to the consistency of whitewash. To one peck of limestone, 2 quarts [2 L] of crude carbolic acid, 4 pounds [2 kg] of sulfur, and 4 quarts [3.8 L] of soft soap are added. The trunks of the trees should be painted with this wash in late autumn.

In may be that some suckers have started around the base of your trees since the last trimming. Cut every one of these out before snow comes.

Late in the fall, plow a furrow down through the orchard between every two rows of trees if the ground is apt to be wet. The trees will do a great deal better for this surface drainage. Also perhaps some tile drains are needed underground.

County demonstration orchards are showing good results. The cost of pruning, spraying, and managing the Nicholas orchard of twenty-nine trees has been 43 cents per tree.

HINTS *to* POTATO GROWERS

1. A loose, rich, gravelly, or sandy loam soil is desirable for potatoes.
2. Manure should be applied to the crop that precedes rather than to the potato crop.
3. A clover, alfalfa, black-eyed pea, or soybean sod, plowed under in the fall, will make a good potato seedbed. Measure the depth of the furrow to see that it is 8 inches [20 cm] or more deep.
4. Like produces like. Hill-selected seed potatoes should be more productive than those from unselected plants.
5. If potatoes are sprouted in the light before planting, it will hasten growth. Sprouts should be ¼ inch [6 mm] long.
6. Treat all seed potatoes for scab before planting. Here is the most approved method of treating them to prevent a scabby crop: soak the whole seed for 2 hours in a mixture of 1 cup [240 ml] of formalin (often called formaldehyde) and 15 gallons [57 L] of cold water; dry the seed, cut, and plant in ground that has not recently grown potatoes.
7. Do not plant late potatoes too early. Large potatoes planted early must be checked during the dry summer and fall to mature a full crop.

8. Never follow potatoes with potatoes. Rotate crops.
9. A well-prepared seedbed is firm and in good tilth. Preparation before planting is half the battle.
10. Spray plants with Bordeaux mixture at least four times at two-week intervals after the potatoes are well up.
11. Add arsenate of lead to the mixture to destroy bugs. Don't wait until the bugs begin work. Get the arsenate of lead on the plants first.
12. A crop of 200 bushels of potatoes requires 660 tons of water—equivalent to 6 inches [15 cm] of rainfall.
13. Destroy the weeds. Harrow the soil before the plants appear above ground. Such harrowing kills millions of sprouting weeds and prevents much future work. The best harrow to use for this purpose is a spike-toothed instrument, for with it there is practically no danger of harming the potato sprouts.
14. Make it a business to push the potatoes. Do not allow the potatoes to push you. Cultivate them six or seven times during the season.
15. When growing potatoes on irrigated land, the following things are essential: a carefully leveled piece of ground with a fall of not more than 18 inches [46 cm] per one hundred inches, plenty of water, and good drainage for surplus water.
16. In growing potatoes in the Great Plains or dry-land section, every method should be used to store up moisture and conserve the supply.

HOW *to* TRAP *the* RACCOON *and* OPOSSUM

The raccoon and opossum are quite hard to capture, especially the raccoon. They are found in wooded country, usually not far from water. The raccoon lives throughout North America, but the opossum does not range very far north. Generally speaking, the animals are most numerous in swamp and marshland. The flesh is good to eat and can be sold in many cities. The pelt hunter can make more money trapping the animals when there is a market for both the skins and the carcasses.

The beginner must pay attention to the traps that are used for taking the raccoon. The animal is large and strong and often pulls out of holds that would prove effective for other furbearers. Unless fastenings can be put in deep water, it is best not to employ stakes. Wire the chains to rocks or logs instead. The rocks or logs should weigh at least 25 pounds [11 kg]. A captured raccoon can drag these but a short distance and does not get a straight pull nor much of an opportunity to work its leg from between the jaws.

Nothing smaller than a No. 1½ trap should be put out. Catches can be made with No. 1 traps, but it is best not to take any chances of losing the skins. If stakes are utilized, do not use soft wood. The raccoon has strong, sharp teeth and uses them to advantage. It is not unusual to have them gnaw the stakes and escape with the trap. This causes needless suffering—a thing the trapper should prevent at all times.

The raccoon, like its larger brother, the bear, has a keen sense of smell and always seems hungry. A good bait is almost a necessity in trapping the animal. Among the baits used by professionals are honey, fresh-smoked fish, canned corn, clams, etc.

When making sets, the trapper should be careful to hide all boot marks or other evidence of his presence. Unless this is done, but few skins will reward him. Sets for the raccoon give best results when placed in water. A good method is to build V-shaped pens of rocks near shore, place a bait in the back part, and guard it with one or more traps. If honey or canned fish is used for bait, it must be above the water. If a small perch or sucker is used, stake it just below the surface so it can easily be seen. The idea is to give a natural appearance to the fish.

Do not overlook partly submerged hollow logs anchored along a stream. Place traps at each entrance. If the water is too deep, build a base of mud, stones, or other material for each set; when the water is shallow, make an excavation for the trap. No decoy is needed for a set of this kind, for the first animal in passing will try to enter the log.

Search for raccoon tracks where small streams enter larger ones. The tracks resemble the imprints of a small baby's foot. Once seen, they are not easily forgotten. As a rule, in places like those just mentioned, concealed traps will prove very effective. In case there are no distinct animal paths, use a bait and arrange the set so that the animal in investigating is bound to be taken.

Raccoon Tracks

In shallow water, not more than 10 feet [3 m] from shore, open clams and arrange them in a pile so that the top is slightly above the surface. Surround the pile with traps. For this plan the water must be reasonably clear and the set must be hidden by grass, moss, or similar material from the bottom of the stream or pond. Make a trail of canned salmon leading to the spot.

The raccoon often travels under shelving banks, and it is there that some of the largest catches can be made. When such a location has been found, conceal the trap where the tracks enter the water. Land sets like those used in taking skunks and civets may frequently be of value if care is taken in the arrangement. Traps can be concealed in trails leading into cornfields. Raccoons never fail to investigate any bright object in the water. Knowing this, pelt hunters often place a piece of bright tin near their traps, so that when a raccoon approaches to investigate he is taken.

In steep banks, dig pockets about 18 inches [46 cm] deep and 4 to 5 inches [10 to 12 cm] in diameter. Have these so that the entrances are in water but the back parts above it. Put bait in these excavations. Smoked fish is good bait because it gives off an odor that the animals can detect and locate. Conceal these sets with moss or water-soaked leaves.

The opossum is not so hard to trap as the raccoon. One of the best ways is to hang sardines about 1 foot [30.5 cm] from the ground in bushes and conceal traps under them, covering with something natural to the place. The opossum can smell the sardines for long distances.

Good places to trap opossums can be found along streams. They travel in such places, generally following a trail in which sets may be made. If there are no distinct paths, use rabbit, fowl, or muskrat for bait, suspending the bait above the trap on a stake. If fowl is used, pluck and leave the feathers near. Scatter particles of canned fish where the opossum travels in trails leading to the traps. Sets should not be too close together. Several trails 100 feet [30.5 m] long will accommodate about two or three traps each. If opossums are numerous, it is possible to capture a half dozen during a night with the method just explained.

Opossum Tracks

SKINNING, STRETCHING, *and* DRYING PELTS

Next to catching fur-bearing animals, too much stress cannot be placed on the importance of properly skinning and stretching, so that the best prices may be obtained. Skins improperly removed and stretched will often bring 10 to 30 percent less than they would if properly handled. If well stretched, you can tell better what size they are, and therefore grade them properly, enabling you to know whether you are receiving full market value.

Fig. 1

Skin animals as soon as possible after taking them from the traps, but not until their fur or hair is dry. To skin an animal cased, cut from the toes of one hind leg. Make no other cuts. Be careful not to cut the skin or leave any more fat or meat on the skin than can be helped. Skin down over the head and eyes, skinning even the nose. Skin very carefully over the ears and eyes. Cut tails from muskrats and opossums only. Skin the feet out of valuable animals.

To skin open, make additional cuts from one front foot to the other and from the base of the tail to the under lip. Stretch the skin in its natural shape on a board, tacking it fast at the edges.

After skinning, stretch as soon as possible in its natural position on the board. Stretch fully, but not too much. Keep the pelt-side out. It is not necessary to turn skins, but sometimes mink and fox skins are turned. Turn in 12 to 24 hours. Pelts that are skinned cased should be tacked at the large end of the board. When the skins are dry enough to hold their shape well, remove them from the board.

The boards for cased skins should be planed smooth, with beveled edges, and taper gradually. The thickness should be ¼ to ¾ inch [6 mm to 2 cm], according to the skin. Shingles are fine for making muskrat boards. For convenience in handling, the boards should be a few inches longer than the pelts.

After the skins are stretched, they should be placed in a cool, shady place (never in the sun or near a fire) to dry. When partly dry, scrape with a dull knife to get the fat and meat off, but do not go too deep. Freezing does not hurt skins. When shipping, pack flat, fur side to fur side, or pelt side to pelt side. Always put a tag on the inside, and insure your furs, if shipping by parcel post.

1. Muskrat board.   2. Plain mink board and key.   3. Three-piece board.

In preparing opossum skins for market, do not split the pelts open, but case them pelt-side out. Scrape off all superfluous meat and fat and chop off the tail, as this is worthless and only adds to the weight of the skin. The pelts need to be dried only long enough to hold their shape.

The ARCH PEST–RATS

At one time our premises were so overrun with rats that we sustained quite a loss from their devastation. A plan for their destruction was devised, as follows:

THE KETTLE TRAP

Filling an iron kettle three-fourths full of barn sweepings, corn-cobs, and a little mixed grain, we set it in an empty stall in the horse stable where the rats seemed to predominate most, and left it this way for some time, keeping plenty of grain in with the rubbish as an enticement for the rats. We laid several boards sloping from the kettle to the floor, so that the rat could easily run up and down and into the kettle.

At the end of about 2 weeks, or when we thought a great number of rats had become accustomed to frequent the kettle, we emptied the kettle of its rubbish contents and filled it three-fourths full of water and covered the water about 1 inch [2.5 cm] or more with light chaff, leaving no water exposed. (If water remains entirely undisturbed, the chaff will not sink overnight.) On the chaff we scattered a little light grain. There was something going on that night! The rats had a party or something; at any rate, the next morning when we went to fishing we scooped out about a half bushel of rats, big and little. The next morning our haul was not quite so large, but we got quite a number; and so on until the rats either got wise or there were no more rats. If we did not get all, we at least got a large majority of them.

THE TAR CURE

At another time when rats were getting altogether too plentiful, we caught a rat in a box trap. This rat we let run into a grain bag and there we caught it by the nape of its neck, guarding carefully against being bitten; then we let all but the head and neck come out of the bag and painted all over the exposed parts of the rat thoroughly with tar and let the rodent go. We had heard that doing this to one rat and letting it go would clean the premises of all other rats, as they object to the smell of tar, or are frightened at the strange appearance of one of their party. It seemed to work in our case and work well. We had no trouble with rats for several years after that. Lonesome, heartbroken, or what, I don't know; but one morning shortly after we had tarred this rat we caught the same fellow again in the same trap we had caught it in before. However, this time we did not let it go.

STEEL TRAPS

It seems that in no other place are rats so hard to catch as in the cellar. Located there they seem to be able to evade all traps and trapping. But I found a way to get Mr. Rat in the cellar. I set a steel trap and put it in a shallow, discarded bread pan and covered the trap completely with wheat bran; the bran, being light, did not spring the trap nor hinder the working of it. Over and about the trap on the bran I scattered a few bread crumbs or meat scraps. This method has never failed me in getting rats in the cellar, although it has when tried in other places. The bran and foregoing baits differed so much from the edibles the rats in the cellar were accustomed to diet on that they jumped for the chance of a change and consequently were easily caught in this manner.

I have found that rats often gain entrance to a cellar through the cellar drain, and for this reason the outlet to the drain should be screened so that no rats can enter.

Chloride of lime, if generously sprinkled over the runways of rats, will also clear the premises of the pests. It gets into their nostrils and burns their feet. Rather than brave many repetitions of it they leave the premises.

Prevention is sometimes better than cure. Where possible to do it, use concrete for floors, foundations, etc. The additional cost of thus making buildings ratproof is slight as compared to the advantages. With cement, even an old cellar may be made inaccessible to these pests.

Rats are expensive; they are destroyers of property. They are a menace to health, carrying in their fur disease germs; they are transmitters of plagues, a general nuisance—out with the rat!

FIRE
PROTECTION
that
PROTECTS

"Well, you saved the barn, anyway," I said, consolingly.

"Yes—by sheer good luck," grunted the owner of Chesapeake Farms, picking a dented fire pail from the cinders. "The wind happened to be blowing the other way; that was all."

"Couldn't you get a fire stream on it? I thought you had a good water supply!"

"I thought so, too. I had a pressure pit under my shop and a gravity tank over it, on a high iron tower. But the fire started in the shop and burst through the roof before we discovered it. In 2 minutes the iron supports of the tower were red hot and crumpled up—there the thing lies."

He pointed to what looked like the blackened, tangled framework of a wrecked zeppelin airship. "Of course, when the tower tank fell, it landed on the pressure tank, smashing the valves off that; my gasoline engine and pump were in the shop, too; the fire buckets had been carried off to slop the hogs— and there you are!"

SIMPLE THINGS TO REMEMBER

Now, all this isn't an argument against fire protection; it's precisely the opposite. My friend did *not* have a good fire system, and so he lost several thousand dollars' worth of farm buildings, with all their contents. Iron is far less fireproof than stout timbers; it bends like wax when hot and should never be used for a tank tower, unless set away off by itself. The pressure tank should have been buried in the ground. The pumping engine ought to have been in a small, isolated building. And so on.

In these busy days, a farm fire is as much a national calamity as the destruction of a steel mill or a shipbuilding plant, and it's a patriotic duty for all of us to protect our farm buildings carefully.

Common whitewash, with a little salt added, makes the best possible fireproof paint. And Pacific Coast redwood scarcely burns at all. In the San Francisco fire of 1906, some redwood houses actually stopped the progress of the flames.

In a large, connected mass of farm buildings, fire partitions can be run up, so that a fire can be kept from spreading. These partitions should of course cut right through the roofs and frame walls and can be made of brick, cement block, hollow tile, or metal lath plastered with cement. All doors through such partitions must be tightly covered with tin on both sides. Fire extinguishers are good thing to have handy.

YOUR WATER SYSTEM

If you have a water system, it should keep pressure enough to throw a good stream against the highest point of any building. A pressure at the ground of 30 pounds [13.5 kg] will shoot the water about 40 feet [12 m] in the air, using 2½-inch [6 cm] fire hose.

If you have only the ordinary garden hose, a much greater pressure is necessary; the company you buy your water tank from will figure it all out for you.

But the best possible fire protection is a sprinkler system; there are dozens of good sorts on the market, and practically every factory, large or small, is equipped with one.

Then, there are all sorts of things you can do to keep fires from starting. When I visit an old farmhouse I always examine the chimneys very carefully; nine times out of ten I find gaping holes right through the brickwork, just under the roof! And then there's the danger of spontaneous combustion from greasy rags, the danger from lightning, etc.

HOW *to* EMPTY *a* BARREL

Those who usually think it hard will find this way easy: Wooden barrels containing 50 gallons [190 L] of fluid, weighing 600 pounds [272 kg] and upward, are very hard to handle except by rolling. In order to withdraw the contents by means of a faucet, it is necessary that the barrel be placed on a raised platform. By using the following method of drawing off part of the contents, the barrel can be handled much easier.

With the barrel lying on its side on the ground, place a stick of wood under one end to cause that end to be slightly higher than the other, as shown in the barrel marked "A" in the picture. Bore a ⅜-inch [1 cm] hole in the head at the uppermost point next to the chime. About 6 inches [15 cm] to the right bore another hole with a 1-inch [2.5 cm] bit, which makes the right-size hole for a piece of ¾-inch [2 cm] pipe. Then screw a piece of pipe 4 to 6 inches [10 to 15 cm] long in the second hole, and fit a plug into the smaller one. To withdraw some of the contents, place a pail under the pipe and roll the barrel to the right until the flow starts, then withdraw the small plug and admit the air so the flow will be increased.

The Plan in Operation

To check flow, insert the plug in the air hole and roll the barrel back to the left. If the barrel is chinked in such a position that the pipe is at the top of head, it will not even need a plug in the pipe.

This idea can be readily used when creosote, kerosene, sheep dip, molasses, oil, spray material, or anything of a like nature is received in barrels. When the barrel has been about half emptied, it may be fitted with a faucet and then rolled upon a raised platform.

EASY-TO-BUILD ICEHOUSE

Some farmers are thoughtful enough in winter to look ahead and store up ice for the summer, and they truly enjoy the fruits of their labors. On the farm, ice is a necessity and also a luxury.

Ice is used for a large number of purposes. It deserves a place on every farm. Icehouses can be built with only a small outlay of money, as skilled labor is not necessary for the construction if care is used, and in almost every locality there will be found running water that can be dammed up into a small ice pond.

The icehouse shown here is made of 12-inch- [30.5-cm-] thick hollow building tile, with a 1-inch [2.5-cm] cement coating inside. It keeps ice with but little cost during the summer months, and is permanent. In size it is 14 by 20 feet [4.25 by 6 m], and the capacity is 50 tons [45 metric tons]. In many places the materials needed for this structure can be had for approximately $175. The door is dirt; the footings for the tile walls are concrete, set down deep so as to be under the frost, and proper drainage is provided. Mix the concrete 1:3:5. Lay the tile blocks in the wall with a lime and cement mortar, using only enough lime to make the mortar sticky. Build the walls true and in perfect line to prevent any possibility of cracking or bulging. The walls go up 10 feet [3 m].

At the top, bolt the wood sill every 5 feet [1.5 m] to the wall, so that the rafters can be securely spiked in place. This will make a solid connection between the tile and the frame roof and prevent the danger of wind blowing the roof off. The 2×4 [5×10 cm] rafters are set 2 feet [61 cm] on-center and at one-third pitch, with a crosstie on every set of rafters so as to form a support for the ice carrier track.

The gable ends of the icehouse are of frame and are fitted with ventilators. The rafters are covered with tight boards, preferably sheathing lumber, laid with tight, smooth joints to make a trim and solid foundation for the roofing felt.

At one end of the house the door extends from the ground up to the plate. This is built in sections for convenience and is best made double thick with building paper between the layers of boards. Cut some short lengths of planks for loose leaves that can be set in across the doorway so that the ice and the sawdust floor will not crowd against the door.

Materials Needed:

10 barrels cement for wall footings, etc.
650 hollow tile blocks for wane
Two 14-foot [4.25-m] 2×12s [5×30.5 cm], for plates
Two 20-foot [6-m] 2×12s [5×30.5 cm], for plates
Twenty-two 10-foot [3-m] 2×4s [5×10 cm], for roof rafters
Ten 8-foot [2.5-m] 2×4s [5×10 cm], for cross-ties
Two 10-foot [3-m] 2×12s [5×30.5 cm], for doorframes
630 feet [192 m] of roof sheathing
100 feet [30.5 m] of 6-inch [15-cm] flooring lumber for
*   doors and gables*
4½ squares three-ply roofing felt

The location of an icehouse is important. Often it is placed near the pond or stream where the ice is secured, thus doing away with a long haul for the ice. This is handy at filling time but not very convenient in the summer when ice is needed and all the help is busy in the fields. Have the icehouse near the dairy house and not far from the house. Then the farmhands will have time to fill the refrigerators and iceboxes during the noon hour, without hitching up a team. So much the better if the building is shaded and open to the north.

AN ALARM BELL

B-R-R-R-R-R!! The electric bell interrupted my delightful dream. I sprang out of bed, raised the south window, and
snatched up my shotgun (loaded with powder only).

"It's someone at the chicken house," whispered my wife.
"Quick! Before they break in and get some of our nice broilers!"

Bang!! Bang!!! Then, hastily pulling on some clothes, I
ran out into the moonlight to investigate. The big padlock lay
wrenched to bits beside the open door; the burglar had vanished,
leaving a string of very hasty footprints in the mud. Something
else, too, he had left—a big phosphate bag, holding eleven of
neighbor Miller's barred Plymouth Rocks!

Now, you may not be so lucky as neighbor Miller, when the
chicken thief, or any other thief, for that matter, comes around; so
a bit of practical burglar insurance in the shape of a burglar alarm
may be a very good thing. Here's the general arrangement—an
electric bell in your bedroom, a wire from this to a little switch
near at hand (so you can cut off the current in the daytime),
a wire over to the circuit breaker on the chicken house door,
then another wire back to the house, thence to the battery, and
so on to the bell again.

Two dry batteries will do the work; the discarded cells from
a gasoline engine have enough current to run an electric bell
for a month or two. The bell is the ordinary electric doorbell;
the switch can be made using a bit of spring brass or copper, on
a wooden block. Iron telephone wire or barbed wire will do for
the outdoors line, if it is wrapped around insulators made from
bottle necks and mounted on trees, poles or fence posts. Only
one wire need be thus insulated; the chicken-yard fence will
answer for the other. Inside the buildings, insulated electric-bell
wire had best be used, for one side of the circuit at any rate.

At the door you can put in a regular ready-made burglar-alarm connection, but I made something just as good myself. I first took off the hinge and reamed out one of the screw holes a little larger than a lead pencil; then, in line with this, I bored a ¾-inch [2-cm] hole in the doorframe, about 1 inch [2.5 cm] deep; changing to a ⅜-inch [1-cm] bit, I bored 1 inch [2.5 cm] or so deeper. Then I made my fixture with a hardwood spindle, a coiled spring, two washers, and a brad driven through the spindle. One wire is fastened to the coiled spring; the other is underneath the hinge plate. Now, when the door is opened, the spring forces the spindle out until the metal washer touches the metal hinge; this completes the circuit, and the bell rings. Brass or copper washers and hinges are best; rusty iron will not make a good connection.

Other doors can easily be attached by running two wires and wrapping the ends of these firmly around the outgoing and incoming main wires, respectively. A double-hung window usually has a special fixture, rather too difficult for you to make, although you can get around this by setting a door fixture in the sill and letting the sash shut down on it. A small movable chicken coop can be protected in the same way, using the loose floor platform as the windowsill.

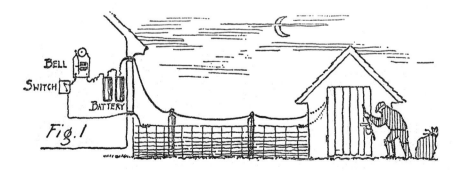

BELL
SWITCH
BATTERY
Fig. 1

PART III

In the

KITCHEN

HOMEMADE WARMING OVEN

If one will try to procure a square cracker tin from the grocer, he can provide a very excellent warming oven for dishes. A button of iron is riveted to each corner to raise the tin a bit above the stove top while it sits on it. Two sheet-iron slides are fitted to the interior of the tin so that dishes can be inserted at once, and the work is done.

WARNING *for* BAKE OVEN

A notice on an oven door—even a sign that simply says "look in"—helps call the cook's attention to what she has put in the oven and is in danger of forgetting as she bustles about doing other work meanwhile. This notice on the oven door, in big letters, is sure to keep the oven's contents in her mind as she passes back and forth through the kitchen.

HEAT BOX *for* RAISING BREAD

This arrangement exactly reverses the operation of a refriger-
ator. A tall wooden box or cabinet with a shelf in it should be
obtained, allowing a pan of dough to placed on top of the shelf,
and a hot water jug put at the bottom. The heat rises all night
long from the jug and surrounds the pan of dough. The box
should be perfectly tight, woolen strip being used for the door
to shut against. In very cold weather a blanket can be thrown
over the box at night.

THE SIMPLE ART
of
MAKING CHEESE

So much has been written about the use of skim milk, butter-milk, and whey for feeding calves and pigs that it seems to have created an impression that those dairy products are not suitable for human consumption. It all depends on how they are used, to be sure. All those products, as well as whole milk, can be made into different kinds of cheese, all of which are palatable and have a high food value.

THE CHEESE IN CONGRESS

Making cheese from sour skimmed milk or buttermilk is not a difficult thing to do—at least not difficult enough to neces-sitate the government's expenditure on agents to go about through the country telling the people on the farms how to make cheese (a plan that was seriously considered in Congress). One of the congressmen, a practical farmer, remarked that there is not a ten-year-old farmer's girl in the country who does not know how to make cheese from sour milk by placing the sour milk—or clabber—in a cloth bag and hanging the bag on a hook to drain overnight, removing the curd from the bag the next morning and seasoning with salt, after which the cheese is ready for use.

Not much needs to be added to the good common sense and plain directions of this western congressman. He did not sug-gest, however, that the milk would be curdled sooner if heated to about 100 degrees F [35 degrees C], although he did prove that he knew all about the process including eating the product, by saying that the quality of the cheese is improved by the addition of cream or butter.

CHEESE FROM WHEY

In some parts of the country a kind of cheese is made by evaporating whey. The whey, not too sour, is boiled in a vessel, using a slow fire and being careful not to burn or scorch it. As the albuminous protein matters are coagulated, they are removed from the kettle and the boiling continued until the whole mass becomes syrupy. The albuminous matter is then returned to the kettle, the vessel removed from the fire, and the contents mixed rapidly to form a thick mush. Some cream is added and the material pressed into molds, after which it is ready to be eaten or sold.

CREAM CHEESE

Cream cheese can be made from whole milk, and the work of making it is often a welcome change from butter making. To make a batch of cheese weighing 3 to 4 pounds [1.4 to 1.8 kg], heat 5 gallons [19 L] of milk to 100 degrees F [35 degrees C], stir into it 1 tablespoon of salt; dissolve half of a rennet tablet in a little cold water and add this to the milk. Remove the milk from the fire and allow it to stand for 10 minutes, when it will become curdled. The curd is carefully cut lengthwise and crosswise with a sharp, flat stick and kept standing a little while longer to get the whey to separate naturally. Having dipped off most of the whey, the remainder is poured slowly into a strainer to drain out still more whey, cut into inch cubes, and salted until they seem a trifle too salty. It can be pressed in a round or square hoop, as you may choose, though there will be less waste in the round. Such a hoop may be a round tin can or one of wood with a follower—a pressure plate—in both ends of the hoop to hold the curd, and must stand on a draining board. It must be lined with a cloth long enough to twist over the top of the curd before the top follower is put in.

The curd is pressed slightly for 8 to 12 hours by placing a weight on the top follower. During pressing, remove the cheese occasionally to trim it and shape it and get its final cap on. Stand the pressed, green cheese in a cool place until dry and then dip it quickly into hot beef tallow or paraffin. Keep it in a cool, dry room not warmer than 60 degrees F [15 degrees C].

EASILY PREPARED DISHES

DOCK GREENS

The curly leaf dock that grows wild in undisturbed fence corners is even more wholesome and palatable than spinach.

CHICKEN PILAU

The rooster is the foundation for the famous dish of chicken and rice that the French call *pilau*. Make it by boiling the chicken on the stove top until the meat comes off the bones. Season well, and add enough rice to thicken the stew. Cook until the rice is soft. The stew should be thick enough to eat with a fork. Serve with a green salad or stewed fruit for a complete meal.

SAUSAGE AND HOMINY

When you want a hearty dish for supper, spread cold boiled hominy—coarsely ground corn—over the bottom of a baking dish, moisten it with milk, season with a little salt and pepper, and slice over it some sausage. Bake in a rather hot oven for 20 minutes, or until the sausage is well cooked, and serve hot.

POTATO PANCAKES

1 cup [240 ml] riced or grated potatoes, ½ teaspoon salt, 1 egg (beaten), 1 tablespoon flour, ¼ cup [60 ml] milk. Mix the ingredients in the order given, beat thoroughly, and cook on an aluminum griddle to save fat.

PENNSYLVANIA PEPPER HASH

Pepper hash is a good old Pennsylvania relish that is delicious served with oysters, codfish balls, cold meats, etc. It is easily made and keeps indefinitely. The proportions vary, but 1 red and 1 green sweet pepper to about 1 pound [455 g] cabbage is a good rule; more cabbage may be added if desired. Also, 1 teaspoon each celery and mustard seed, 2 teaspoons sugar, a tiny piece of horse radish, and enough vinegar to cover the mixture. Remove the seeds from the peppers, cut them and the cabbage into shreds, then chop fine. Sprinkle with 1 tablespoon salt, stand aside for 1 hour, then drain all water away. Add other ingredients, mix, then pack down hard in a crock or glass jar, pour on the vinegar, diluted if necessary, and cover, but do not seal. In summertime, a few nasturtium leaves may take the place of the horseradish, which will prevent fermentation. During the winter, if sweet peppers are not available, one or two cucumber pickles and a red pepper pod (containing the seeds) will make a relish that tastes as good as it looks.

SAVING COOKED CEREALS

No cooked cereal needs to be wasted. Even 1 teaspoon can be added to the batter for hot cakes or muffins. A small quantity of oatmeal may be used to thicken a soup.

GRAPE FOAM

Grape foam, which consists simply of the white of 1 egg beaten stiff and added to 2 tablespoons grape juice, is a delightful drink for an invalid. It will quench the thirst of fever and prove nutritious as well. Orange albumen is prepared in the same way, using orange juice instead of grape juice.

ITALIAN POLENTA

Polenta, an Italian way of serving cornmeal, makes almost a meal in itself. Put slices of cold mush (cornmeal pudding) in a baking dish and add 1 cup [140 g] of sliced onions that have been fried in ham or bacon fat. Add 2 cups [480 ml] canned tomatoes and cover all with 1 cup [125 g] grated cheese; bake until the cheese is melted and slightly browned.

BUCKWHEAT SPICE CAKES

Buckwheat spice cakes offer an excellent way in which to use up your surplus buckwheat flour. Use ¾ cup [150 g] sugar, 3 table-spoons fat, 1 egg, 1 cup [240 ml] sweet milk, 1 cup [120 g] buckwheat flour, 2 teaspoons baking powder, ½ teaspoon salt, 1 tablespoon cinnamon, ¼ teaspoon cloves, 1 tablespoon vanilla. Mix like an ordinary cake. Bake in muffin tins in a moderately hot oven.

GOLDENROD EGGS

Goldenrod eggs make a wholesome and attractive dish. Boil until hard as many eggs as are needed. Cut the whites in narrow strips and place on slices of toast. Pour over this a white sauce, and sprinkle the egg yolks, pressed through a potato ricer or finely grated, over the top. White sauce is made by rubbing to a paste 2 tablespoons each flour and butter, then pour on gradually (stirring all the while) 1 cup [240 ml] hot milk. Bring to boiling point and season with salt and white or cayenne pepper.

SPINACH PROPERLY COOKED

Spinach, "the broom of the stomach," ought to be on the spring dinner table several times a week, and asparagus the other days. Wash well some young spinach leaves, shake off the water, and put in a hot stew pan. Shake and stir until the juice of the leaves comes out and they seem to be melted, then cover and cook slowly for 20 minutes. Cooked thus, all of the valuable vegetable salts are retained. It may be eaten with butter or vinegar, as preferred. Slices of hard-boiled egg add to its appearance and flavor.

ROASTING IN PAPER

The French have a way of making a year-old fowl as tender as one that is half the age by wrapping it in brown paper before it is put into the oven and allowing it to cook in this envelope until it is nearly done. The paper retains the juices and allows the fowl to cook slowly and evenly and grow tender before the outside is browned. At the last the paper is removed long enough to bring the surface of the fowl to the desired color. Young mutton can be brought to lamblike tenderness in the same way, and roast veal may be cooked thoroughly without the hard outer crust that sometimes spoils this meat when roasted.

BARLEY CEREAL

Barley is one of the grains recommended as an occasional substitute for wheat, and though it has hitherto been principally used for soups and gruel for babies and invalids, it is nourishing and palatable; it builds muscle and makes bone. To use as a cereal, wash ¼ cup [50 g] barley and put it into a double boiler with 4.5 cups [1 L] cold water and 1 teaspoon salt. Do not stir but let it cook for 2 hours. It should swell to four times its original size and be like jelly. Raisins can be cooked with it, if desired. Serve with cream.

SKIM MILK BREAD

Skimmed milk used in bread in place of water adds as much protein to a loaf of bread as there is in one egg. It gives a softness of texture to bread that adds particularly to the palatability of graham or bran bread.

POTATO MUFFINS

Potato and cornmeal muffins are delicious. To make them, use 2 tablespoons fat, 1 tablespoon sugar, 1 egg well beaten, 1 cup [240 ml] milk, 1 cup [325 g] mashed potatoes, 1 cup [150 g] cornmeal, 4 teaspoons baking powder, 1 teaspoon salt. Mix in the order given, put in muffin pans, and bake 40 minutes in a hot oven.

MIXED BREAKFAST FOOD

A delicious and nutritious breakfast food can be easily made by mixing together equal parts bran, oatmeal, and cornmeal. A few raisins chopped finely and mixed thoroughly into the porridge just before removing from the fire will add greatly to the flavor and no sugar will be needed. Stir the cereal gradually into boiling water and allow it to cook slowly for 15 or 20 minutes. If there is any left over, it can be fried like cold mush.

BOILED RICE

Boiled rice should be prepared thus: Wash 1 cup [180 g] rice several times. Then drop it slowly into 4.5 cups [1 L] boiling water, salt to taste, boil for 15 minutes, and then remove from burner. Cover and place on the back of the stove where it will finish swelling without burning. Do not stir. To serve, arrange the rice in a ring on a hot dish and place in the center any meat hash, stew, creamed fish or chicken. Nothing more is needed for dinner but a dessert of stewed fruit or a green salad.

MOCK CHERRY PIE

A good substitute for the cherry pie that you will want to serve on Washington's birthday is made by cooking together 1 cup [140 g] cranberries, halved, ½ cup [80 g] seeded raisins, ½ cup [120 ml] water, and a dust of flour. When the cranberries are done, add 1 teaspoon vanilla and sugar to taste. Line a pie plate with good pastry, put in the filling, add the top crust, and bake.

BOILED SOYBEANS

Boiled soybeans form the basis of many nourishing dishes. As thorough cooking is essential, follow these directions: Soak 2 cups [450 g] dried beans for 24 hours. Drain, add boiling water, 1 tablespoon salt, and a pinch of baking soda, and simmer for 4 hours. The beans may also be cooked with 8 ounces [225 g] fat salt pork, cut in cubes.

POPCORN CEREAL

Allow the children to make their own crisp breakfast food. Popcorn, fresh, warm, and crisp, smothered in whole milk and dusted with sugar, is delicious. The children will enjoy this food because they can prepare it and eat it perfectly fresh. Popcorn pudding will also please the children. Run 2 cups [16 g] popped popcorn through the food grinder, add 2 cups [480 ml] milk, 1 tablespoon butter, ½ cup [100 g] sugar, 1 egg lightly beaten, and a sprinkle of salt. Bake 30 minutes and serve warm.

BROWN SUGAR PICKLES

If a sugar shortage prevents you from making your favorite kind of sweet pickles, try these: Cut one dozen plain sour pickles in slices ¼ inch [6 mm] thick. Place in layers in a jar, sprinkling each layer with brown sugar, a few cloves, and a little stick of cinnamon. Stir or shake every day for one week, when they will be ready to use. One pound [455 g] brown sugar is sufficient for a dozen large pickles, and the syrup formed in the first jar may be used to start a second jar of pickles.

APPLES AND DATES

Apples and dates make a good combination and require no sugar. To prepare them, steam 6 cups [1.4 L] sliced apples and the grated peel of 1 lemon with ½ cup [120 ml] water until tender in a covered pan. Add ½ cup [90 g] chopped dates, simmer the fruits together for 6 minutes, and serve cold.

CORN FRITTERS

Corn fritters are good for a change. To make them, run through the meat grinder a can of the corn you put up last summer. Add 2 eggs, well beaten, just enough flour to hold the mixture together, and salt to taste. Beat thoroughly, then drop by the spoonful into a pan in which there is hot fat to keep them from sticking. Fry a nice brown on each side and serve hot.

REWORKED BUTTER

To make reworked butter, use 1 pound [455 g] butter, 2 cups [480 ml] rich milk, 1 tablespoon gelatin, and salt to taste. Cream the butter, as you would for cake, then dissolve the gelatin in a little of the milk, heat the rest of it, and pour it over the gelatin. When this becomes lukewarm, pour it slowly over the butter and work in a small churn or beat with an eggbeater until well mixed, smooth, and thick. Drop by the spoonful in a shallow dish and keep in a cool place.

BOSTON BROWN BREAD

Boston brown bread is a good standby. To make it, use 1 cup
[140 g] white flour or bread crumbs, 1 cup [150 g] cornmeal,
1 cup [110 g] graham flour, 1 teaspoon salt, ¾ cup [240 g]
molasses, 1¾ cups [420 ml] fresh whole milk, ¾ teaspoon baking
soda, and 1 teaspoon baking powder. Sift the dry ingredients
together, then stir them into the liquids. Fill well-greased cans
two-thirds full, cover tightly, and steam for 4 hours. Good
either hot or cold.

ORANGE MARMALADE

If you have reason to think that your jelly supply will not
last until spring, try this recipe for orange marmalade:
Use 3 oranges and 2 lemons, or for a bitter flavor, 1 large grapefruit, 1 orange, and 1 lemon. Squeeze out all the juice and
put all the skins through a meat grinder. Put both together
and add three times as much water as the entire quantity of juice
and skins. Let stand overnight. The next day, boil for 10 minutes, then measure and add an equal amount of granulated sugar.
Boil for 1½ to 2 hours and put up as any other marmalade.
This quantity makes about 12 jars. If you can get oranges for a
fair price, the marmalade will be very inexpensive and will be
a wholesome and delicious sweet for your table.

MEASURING HARD BUTTER

To measure butter without softening it: If ½ cup [110 g] is
needed, fill 1 cup half full of water, then add pieces of butter
until the cup is full. If 1 cup is wanted, repeat the process.

CORN BREAD WITH RICE

Corn bread with rice is an economical dish, requiring 2 cups [480 ml] sour milk, 1 scant teaspoon baking soda, 2 cups [400 g] boiled rice, 1 cup [150 g] cornmeal, and 1 tablespoon shortening. Combine the ingredients in the order named and bake in a greased pan until firm. Serve from the pan with a spoon.

SCOTCH OAT CRACKERS

Scotch oat crackers are crisp and good as well as cheap. To make, take 2 cups [200 g] rolled oats, ¼ cup [60 ml] milk, ¼ cup [80 g] of molasses, 1½ tablespoons fat, ¼ teaspoon baking soda, one teaspoon salt. Crush or grind the oats in a food mill and mix with the other ingredients. Roll out in a thin sheet and cut into squares. Bake for 20 minutes in a moderate oven.

SOYBEANS WITH TOMATO SAUCE

Soybeans with tomato sauce are good. Prepare the sauce by cooking 1 cup [240 g] stewed and drained tomatoes with a slice of onion, 3 cloves, and a bit of bay leaf and strain. Melt 2 table-spoons butter, add 2 tablespoons flour, and when browned add the tomatoes gradually with ½ teaspoon salt and a little pepper. Cook thoroughly. To 1 cup [225 g] tomato sauce add 2 cups [320 g] boiled beans and reheat.

DELMONICO POTATOES

Delmonico potatoes make a good substitute for meat. Dice cooked potatoes and place them in layers in a buttered baking dish, alternating with layers of thin white sauce and shaved cheese. Use about half as much white sauce and one-fourth as much cheese as potatoes. Sprinkle the top with buttered crumbs and bake in a moderate oven until the top is well browned. Cheese increases the food value of the potatoes, but the latter should be cooked before being combined with the cheese.

LOUISIANA CORN BREAD

Rice improves both plain corn bread and corn bread made with eggs and milk. Use it in both. Corn bread made with eggs and milk is rich in protein and makes a good meal served with a little gravy. This recipe is not extravagant for the cook who has an abundance of eggs and milk. Use 3 eggs, 2 cups [480 ml] milk, 1½ cups [250 g] cold boiled rice, 1½ cups [225 g] cornmeal, 2 tablespoons melted fat, 1 tablespoon salt, and 1 teaspoon of baking powder. Beat the eggs lightly; add the other ingredients in the order named. Beat hard and bake in a shallow greased pan in a hot oven.

CHOCOLATE CORNSTARCH PUDDING

Chocolate cornstarch pudding properly made is almost as good as ice cream. To make it, put into a double boiler 1 cup [240 ml] milk, 1 cup [240 ml] water, ½ cup [100 g] sugar, and 1 ounce [28 g] unsweetened chocolate. When the mixture is boiling, add a level ¼ cup [30 g] cornstarch that has been wet in cold water. Cook until it thickens and the raw taste of cornstarch has disappeared, then add 1 teaspoon vanilla and pour slowly into this the stiffly beaten whites of 3 eggs. Mix thoroughly, then pour into a mold to harden. Serve with a custard sauce made of 3 egg yolks beaten with 3 tablespoons sugar and added to 1½ cups [360 g] hot milk. Cook carefully until it thickens; flavor with 1 teaspoon vanilla and cool. Chocolate should be used largely, as it is both a food and a dessert; it is rich in fat and starches, and a pudding or other dessert in which this product has a part will be welcomed on days when a meat substitute is served as the main dish for dinner.

A TOOTHSOME RICE DISH

A toothsome rice dish is prepared thus: place hot boiled rice on a platter, cover with white sauce, and garnish with sliced hard-boiled eggs and a little finely chopped ham, if you have it.

DRIED VEGETABLES

Dried vegetables can be stored and shipped easily and do not freeze in winter. Small quantities, too trifling for canning, can be saved by drying and will be found palatable and often better than the canned product. In this season of plenty the wise cook will store large quantities of dried vegetables for winter stews and soups. If pressed for time, and sugar happens to be scarce, the cook may put up ample supplies of dried fruit for preserves later, or to be used, after soaking overnight, like fresh fruit for sauces and desserts.

PEACH SOUFFLÉ

This sugar-saving dessert is sure to please. To 4 cups [960 ml] peaches, canned or fresh, use ½ cup [170 g] honey or syrup and 3 eggs. Drain and mash the peaches through a colander; add the honey or syrup and well-beaten egg yolks. Mix thoroughly, then beat the whites of eggs until stiff and fold carefully into the peach mixture. Turn the whole into a greased baking dish and bake in a quick oven set to 375 degrees F [190 degrees C] for 6 minutes.

COTTAGE CHEESE PIE

This pie uses up surplus milk. It requires 1 cup [210 g] cottage cheese, ½ cup [120 ml] maple syrup, ⅔ cup [160 ml] milk, beaten yolks of 2 eggs, 2 tablespoons melted butter, ½ teaspoon vanilla, and salt. Mix the ingredients in the order given, and pour into a pie plate lined with crust. When baked, cool it slightly, cover with meringue, and brown in a slow oven at 325 degrees F [165 degrees C]. For the meringue, use whites of 2 eggs, ¼ cup [60 ml] maple syrup, and ¼ teaspoon vanilla. Beat the egg whites until they are stiff, add the syrup gradually, and then add the vanilla.

PLAIN CORN BREAD

Plain corn bread is inexpensive but good. It is made by sifting together 1 cup [150 g] yellow cornmeal, 1 cup [140 g] all-purpose flour, 2 heaping teaspoons baking powder, and 1 level teaspoon salt. Stir in enough whole milk to make a rather stiff batter and beat it well. Pour the mixture into a well-greased pan and bake in a moderate oven for about 30 minutes. The mixture should be no thicker than pancake batter and if properly made will make a moist, delicious bread.

RECIPES *for* BUSY SUMMER DAYS

STEAMED BROWN BREAD

Steamed brown bread is always good, but especially so in hot weather, as it can be cooked on top of the stove with less heat than would be required if it were baked in the oven. It requires 2 cups [300 g] cornmeal, 2 cups [220 g] graham flour, 2 cups [480 ml] buttermilk, 1 cup [320 g] dark molasses, 2 teaspoons baking powder, 1 teaspoon baking soda, ½ cup [80 g] raisins, and ½ teaspoon salt. Mix in the order given and fill cans or brown bread pots half full and steam for 3 hours. If the cans are placed in a large roaster with other dishes, an entire meal can be cooked at one time over one burner, thus conserving fuel.

QUICK TOMATO SOUP

This tomato soup requires 4 cups [960 ml] water, 2 cups [485 g] canned tomatoes, 1 teaspoon salt, 1 tablespoon chopped onion, bay leaf or parsley, 4 tablespoons butter or butter substitute, and ¼ cup [40 g] barley or potato flour. Mix the water, tomato, and seasonings, heat to boiling, then add butter and flour rubbed to a paste. Boil for 30 minutes; add 2 teaspoons beef extract, then strain and serve.

CREAM OF ASPARAGUS SOUP

Take 2 cups [360 g] of the hard, rejected portions of asparagus, cover them with 2 cups [480 ml] water or stock, cook slowly for 30 minutes; press through a colander, add 2 cups [460 g] cooked rolled oats, 2 cups [480 ml] whole milk, 1 level teaspoon salt, ¼ teaspoon pepper. Stir until the mixture reaches the boiling point. Add 1 tablespoon vegetable oil, strain through a sieve, and serve hot.

CREAM OF CUCUMBER SOUP

This soup is delicious to serve at the beginning of a fish dinner. Peel 2 good-size cucumbers, cut them in slices, cover them with 2 cups [480 ml] cold water, and add 1 grated or chopped onion. Cook slowly for 30 minutes. Add 2 cups [460 g] cooked rolled oats and stir until boiling hot. Add 2 cups [480 ml] whole milk, 1 bay leaf, 1 level teaspoon salt, a grating of nutmeg, and ¼ teaspoon pepper. Stir constantly until it reaches the boiling point, strain through a fine sieve, and serve at once.

DELICIOUS CORN CHOWDER

To make a delicious corn chowder, you'll need 8 ounces [225 g] salt pork, 3 potatoes, 15 ounce [425 g] can of corn, 4 cups [960 ml] milk, 6 pilot crackers, 1 teaspoon beef extract, and salt and pepper to taste. Cut the pork in small cubes and fry until light brown. Parboil the potatoes, cut in cubes, return to the pot, and cover with boiling water. Add the corn, and cook until the potatoes are tender. Add the crackers, which have been soaked in milk, and lastly add the beef extract. Season with salt and pepper and serve very hot.

FISH FLAKE SOUFFLÉ

This soufflé is an appetizing dish, quickly prepared. It requires ¼ cup [60 ml] vegetable oil, 3 tablespoons flour, 2 cups [480 ml] heated milk, 4 eggs, 2 cups [480 ml] flaked fish of any kind, 1 teaspoon salt, ⅛ teaspoon pepper, and 1 teaspoon finely chopped parsley. Make a white sauce by mixing the fat and flour together, then pour on the hot milk, stirring over the heat until it thickens. Beat the egg yolks, add the fish flakes and seasoning, and pour the sauce over them. When cooled off, fold in the egg whites, beaten stiff, and pour into a well-greased glass baking dish. Bake in a moderate oven for 30 minutes and serve at once.

CREAMED FISH CAKES

For creamed fish cakes, rub to a paste 2 tablespoons vegetable oil and 2 tablespoons flour. Add 1 cup [240 ml] milk, and heat in a saucepan until boiling. Season with salt and pepper, stir in 1 cup [240 ml] or one 5 ounce [140 g] can of flaked fish of any kind. Cover and allow it to heat until hot, then pour over slices of toast and serve with baked potatoes.

RED BUNNY

Red bunny is an attractive cheese dish, quickly prepared. Melt 1 tablespoon butter or butter substitute in a saucepan and add 8 ounces [225 g] American cheese cut in thin slices; melt slowly, then add a can of tomato soup, stirring until thoroughly blended and heated. Arrange squares of toasted bread on a platter, pour the mixture over them, and serve hot.

CANNED SALMON

Canned salmon has possibilities not always realized by those accustomed to eating it just as it comes from the tin. It is especially nice combined with rice. Remove all bones and skin from a can of salmon and flake it with a fork. To prepare it, boil 1 cup [180 g] rice; when done stir into it a small can of salmon and serve hot. Accompany this dish with a green salad, corn bread, and stewed fruit.

SALMON SALAD

Remove all bones and skin from a can of salmon and flake it with a fork. Finely dice 1 head of celery and a small bottle of olives stuffed with red peppers. Mix all together with mayonnaise or any preferred salad dressing, and serve on lettuce leaves.

SALMON BOX

This dish is made thus: Line a well-greased ovenproof glass dish or a bread pan with warm steamed rice. Fill the center with cold salmon, flaked and seasoned with salt and pepper. Cover with rice and steam for 1 hour. Transfer to a hot platter and serve with cream or tomato sauce thickened with cornstarch, barley flour, or potato flour.

TOMATO JELLY SALAD

This dish is especially welcome while waiting for tomatoes to ripen. To make it, add an equal amount of hot water to a can of condensed tomato soup. Soften ½ package or 2 tablespoons gelatin in ½ cup [120 ml] cold water. Bring the soup to boiling point, season with salt, pepper, and sugar; remove from the heat, add the softened gelatin, and stir until dissolved. Pour into individual molds, which have been pre-moistened with cold water. When the jelly is cold, remove from the molds and serve on lettuce leaves with mayonnaise dressing.

QUICK MAYONNAISE DRESSING

For a mayonnaise dressing, be sure to use a fresh egg, one with a stiff, firm yolk. Drop the yolk into a deep bowl, add ¼ cup [60 ml] vegetable oil, 1 tablespoon elder vinegar, ½ teaspoon salt, and a dust of cayenne pepper. Whip with an eggbeater until it is thick and heavy. Cover and let stand in a cold place until serving.

KIDNEY BEANS AND SAUSAGES

Kidney beans and sausage makes for an agreeable combination. The canned beans need only reheating, with added seasoning, if desired. Use half-smoked sausages and put into a saucepan, cover with cold water, and bring to boiling, at which point they will be ready to serve. Pile the beans in a shallow dish, arrange the sausages on top, and garnish with watercress.

GRAPE JELLY

This jelly is a refreshing hot-weather dessert easily made by combining one of the quick gelatins with grape juice. Follow the given instructions for making the jelly, using heated grape juice instead of hot water. Whipped cream is a pleasing addition.

JAMBOLAYO

Jambolayo is a delicious dessert, easily made. Fill a glass dish, or individual glasses, a little more than half full with a combination of sliced fruits, bananas, dates, canned peaches, or cherries; add a few chopped nuts, if desired. Make a lemon jelly, using one of the quick gelatins. When cool, pour over the fruit and stand in a cool place to harden before serving.

STRAWBERRY DELIGHT

To make this, cook 2 cups [280 g] berries until soft, then strain and add to the juice enough hot water to make 2 cups. Pour this over a package of quick gelatin, lemon flavor, and, when cool, pour into a ring mold to harden. When set, turn out on a flat dish, fill the center with fresh uncooked berries, and serve with sugar and cream. Raspberries and blackberries are equally good served in this way.

CANNED PINEAPPLE AND DRIED APRICOTS

These make an excellent marmalade that can be used to eke out last year's supply of fruit, or in areas where fresh fruit is unobtainable. Cook 16 ounces [360 g] dried apricots in water enough to cover them, until soft. Then add two 20 ounce [570 g] cans pineapple that has been cut in small pieces. Measure the fruit and add three-quarters as much sugar. Return to the heat and cook until it thickens.

CHRISTMAS SWEETMEATS

There are all sorts of candy substitutes, such as stuffed dates, candied ginger, fruit pastes, and salted nuts. Stuffed prunes are a delicious candy substitute. Wash them thoroughly, take out the seeds, and slip into each one an almond or a peanut and see how eagerly the children will eat them. Dried fruits, such as dates, figs, prunes, and raisins not only have sugar, but are also highly nourishing. Raisins and nuts, if given with moderation, will not prove indigestible.

Eight ounces [230 g] each of dates and nuts run through a grinder, softened with lemon juice, and cut into squares like caramels make a wholesome substitute for candy.

Use more home-salted nuts this Christmas than in previous years. Peanuts, pecans, or almonds, if prepared in olive oil or butter, will not go begging.

To candy orange or grapefruit peel requires the use of some sugar, but less than for its equivalent in candy, and you are using up what would otherwise be thrown away. The following recipes require very little sugar:

PEANUT BARS NO. 1

Put 1 cup [200 g] granulated sugar in a cast iron skillet; stir constantly to avoid burning until it melts to a golden brown. Stir in ½ cup [70 g] broken peanuts and pour at once into a buttered pan.

PEANUT BARS NO. 2

Shell and remove the skins from 4 cups [560 g] roasted peanuts and chop fine. Beat the white of 1 egg until stiff, but not dry, and add gradually 1 cup [200 g] brown sugar, ¼ teaspoon salt, and ½ teaspoon vanilla. Fold the peanuts into the mixture and spread evenly in a buttered shallow pan. Bake in a quick oven set to 375 degrees F [190 degrees C] until well puffed and browned. As soon as taken from oven cut in bars, using a sharp knife.

CHOCOLATE CARAMELS

Mix together 2 cups [400 g] sugar, 2 cups [680 g] extracted honey (or sorghum), 4 ounces [110 g] grated chocolate, ½ cup [120 ml] sweet cream, and 1 tablespoon vanilla extract. Test this mixture often while boiling by dropping a small portion in cold water. When it forms a soft ball, pour about ¼ inch [6 mm] thick on greased tins. Mark in squares just before it hardens.

WALNUT CREAMS

Boil to the hard snap stage 1 cup [125 g] grated chocolate, 1 cup [200 g] brown sugar, 1 cup [340 g] extracted honey (or sorghum), and ½ cup [120 ml] sweet cream. When it hardens on being dropped into water, stir in a piece of butter the size of an egg. Just before removing from the heat, add 2 cups [240 g] finely chopped walnuts, and stir thoroughly. Pour on buttered plates to cool, then cut it into squares.

CRACKER JACK

Boil 1 cup [200 g] brown sugar with 1 cup [340 g] extracted honey (or sorghum) until it hardens when dropped into cold water. Remove from the heat and stir in ½ teaspoon baking soda; when this dissolves, stir in all the popcorn it will take. Spread on greased tins and mark in squares.

WAYS *to* CURE MEAT

The two ways of curing pork are brine curing and dry curing. If brine is properly made, it will keep for a reasonable length of time. If it becomes ropy, it must be poured off and boiled, or a new brine must be made.

A cool cellar is the best place for both methods of curing. Rub the surface of the meat with fine salt and allow it to drain, flesh-side down, for 6 to 12 hours before the meat is cured, either with brine cure or dry cure.

THE BRINE CURE

For every 100 pounds [45 kg] meat, use 8 pounds [3.5 kg] salt, 2½ pounds [1 kg] sugar or syrup, 2 ounces [57 g] potassium nitrate, and 4 gallons [15 L] water. In warm weather, 9 to 10 pounds [4 to 4.5 kg] salt are preferable. All the ingredients are poured into the water and boiled until thoroughly mixed. Then let the brine cool.

Place hams on the bottom of the container, shoulders next, bacon sides and smaller cuts on top. Pour in the brine and be sure it covers the meat thoroughly. In 5 days, pour off the brine and change the meat, placing the top meat on the bottom and the bottom meat on top, after which pour back the brine. Do this again on the tenth and eighteenth days. If the brine becomes ropy, take the meat out and wash it thoroughly, as well as the container. Boil the brine or make new brine, replace the meat in the barrel, and cover with brine. Allow 4 days of cure for each 1 pound [455 g] in a ham or shoulder, and 3 days for each 1 pound [455 g] in bacon sides and small pieces. For example, a 15-pound [7-kg] ham takes 60 days.

THE DRY CURE

This requires more work than brine curing. For every 100 pounds [45 kg] meat, use 7 pounds [3 kg] salt, 2½ pounds [1 kg] sugar, and 2 ounces [57 g] potassium nitrate. Mix all the ingredients thoroughly, rub one-third of the mixture over the meat, and pack the meat away in a box or on a table. The third day, rub on half of the remaining mixture and again pack the meat. The seventh day, rub the remainder of the mixture over the meat and pack it to cure. Allow a day and a half cure for each 1 pound [455 g] of meat. A 20-pound [9-kg] ham will take 30 days to cure.

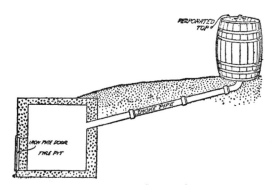

SMOKING

Smoking helps preserve meat that has been cured. It also gives a desirable flavor if the meat is properly smoked with the right kind of fuel. A simple, cheap, and convenient smoking firebox can be made according to the sketch above. The firebox and meat barrel should be at least 6 feet [2 m] apart. Concrete and clay blocks are fine for smokehouses. Wooden houses are not so good, for they are likely to catch fire—and down goes your meat house.

When meat is removed from the brine it should be soaked for about 30 minutes in water before being placed in the smoke-house. When removed from dry cure, it must be washed in lukewarm water. The meat should warm up gradually in the smokehouse and must not get too hot. Green hickory or maple wood is the best for smoking. It takes 36 to 48 hours to smoke meat. Lower and longer smoking will make the meat keep longer. After meat is smoked it should be wrapped in heavy paper and put into muslin sacks to keep insects away. Tie the tops of the sacks tightly.

PREPARING MEATS *for* FAMILY USE

BRINE CURE

When the raw meat is cooled, rub each piece with salt and allow it to drain overnight. Then pack it in a barrel with the hams and shoulders in the bottom, using the strips of bacon to fill in between and put on top. For every 100 pounds [45 kg] of meat, weigh out 8 pounds [3.5 kg] salt, 2 pounds [910 g] brown sugar, and 2 ounces [57 g] potassium nitrate. Dissolve these in 4 gallons [15 L] pure water and, after boiling and thoroughly cooling it, cover the meat with the brine. The bacon should remain in the brine for 4 to 6 weeks, the hams for 6 to 8 weeks.

Meat that is to be smoked should be removed from the brine 2 to 3 days before putting it in the smokehouse. Scrub it clean in tepid water and hang to drain for a day or two before smoking. When smoking, hang it so that two pieces will not touch each other. Make a very slow fire of green hickory, apple, or maple wood, smothered with sawdust of the same. Do not use pine or other resinous wood, as it will give an unpleasant taste. When smoked sufficiently, let the meat air and dry, then wrap in paper and put in heavy muslin or canvas and cover with whitewash.

PLAIN SALT PORK

Rub each piece with fine common salt and pack in a barrel. Let this stand overnight; the next day weigh out 10 pounds [4.5 kg] salt and 2 ounces [57 g] potassium nitrate for every 100 pounds

[45 kg] meat and dissolve in 4 gallons [15 L] boiling water. When cold, pour the brine over the meat, cover, and weight it down. The meat will pack best if cut 6 inches [15 cm] square.

SAUSAGE

For every 55 pounds [25 kg] lean and fat pork chopped very fine, mix together 1 pound [455 g] salt, 6 ounces [170 g] good black pepper, 1 teaspoon cayenne pepper, and a handful of powdered dry sage. Mix these thoroughly through the meat. If you wish to stuff it in skins, clean them thus: Empty the intestines of the pig, turn them inside out, and wash well. Soak them in salt water a day or more. Wash again, cut into convenient lengths, and scrape them on a board with a blunt knife, flat on one side, then on the other, till they are clean and clear. Throw them in clear water and rinse.

Tie up one end of each length, put a quill in the other end and blow them up. If whole and clear they are clean, but if there are thick spots, they should be scraped off. Throw in clean, cold salt water until wanted. To use, put one end over the nozzle of the sausage stuffer and force the meat into them. This can be better done if the meat is first lightly sprinkled with cold water that is worked through it.

Pack the sausage to keep for winter use in stone crocks and run 2 inches [5 cm] of hot lard over it. Sausage that is desired for summer use may be canned, in which case make into small cakes and cook until about two-thirds done, or until all the water is out. Pack in the cans while still cooking, fill them full of hot lard, and seal at once; or it may be stuffed tightly in muslin bags, then the bags rolled in melted paraffin, which should be heated in a large flat pan. When cooked next summer, it will be more delicate if you pour off all the fat after it is fried, pour in a little cream, boil it, and pour over the sausage.

LIVERWURST

Boil the liver with about an equal weight of head meat, including the fat. After it is fairly well done, run it through a food mill while yet warm, season with salt and pepper, and pack it in a crock or into rolls. It should be sliced and fried for use.

PIGS' FEET

Pigs' feet should be thoroughly cleaned. First scrape and wash, then soak in cold water for several hours; scrub and wash well. Split feet and crack in several places. Cover with boiling water, add a little salt, and let them simmer until thoroughly tender, about 4 hours. If you want them pickled, add salt, pepper, and a few whole cloves or allspice to 1 cup [240 ml] good cider vinegar, boil for 1 minute, and pour this over the pigs' feet.

CANNED MEAT

Tenderloin and pork chops can be kept for some time by cutting into serving pieces, frying until rather more than half done, then packing into hot, sterilized jars or cans and running boiling lard over them. Do not put more than enough for one or two meals in one jar, as the meat will not keep after the lard is removed.

DRIED BEEF

Get the tender side of the round out of a good fat beef. For every 20 pounds [9 kg] beef take 2 pounds [910 g] salt, 1 teaspoon potassium nitrate, and 4 ounces [100 g] brown sugar. Mix these well, rolling out any lumps; divide into three equal parts and rub well into the beef for 3 consecutive days. Turn beef daily into the liquor it will make. It should not make much, but what there is rub into and pile on the beef. Rub a little extra salt into the hole cut for the string to hang it by.

At the end of a week, hang in a dry, rather warm place till it stops dripping, then in a cooler, dry place. Do not smoke it; it spoils the flavor. Before flies come in the spring, wrap in paper and put it in a stout bag with a string out to hang by. If it molds some through the summer, scrape and scrub the mold off and always trim the outside before chipping.

BEEF TONGUES

Trim them and lay six or eight tongues into boiling water for 5 minutes. After they are cool, rub them with a mixture of ¼ ounce [7 g] potassium nitrate, two handfuls salt, and 4 ounces [100 g] sugar or a small cupful of molasses. Pack them in an earthen vessel, sprinkling each layer with the mixture and putting a weight on top. Turn them every other day, putting the top one at the bottom, and packing them closely. If there is not enough pickle to cover them, sprinkle lightly with salt. At the end of 2 weeks, hang up to drain. When dry, wrap in paper, put into a bag, tie tightly, and hang in a cool place. If you do not wish to use a whole tongue at once, it does not injure it to be cut in two; but it is best to dip the cut end into boiling water for a moment, in order to seal up the pores.

CORNED BEEF

Wipe every piece with a dry towel. To every 50 pounds [23 kg] allow 1½ ounces [40 g] potassium nitrate, 1½ pounds [680 g] brown sugar, and 9 to 10 gallons [34 to 38 L] water. Mix the sugar and potassium nitrate in the water and add enough good salt to make a brine that will float an egg. Pack the meat in a clean, sweet barrel and pour the brine over, skimming off whatever floats. Cover with a thick cloth; watch the brine carefully for 1 week, skimming it every day. If there's not enough brine to cover, make more until there is.

In about 2 months, drain the brine off and make more in the same way. The meat will keep for 1 year or will be ready for use in 2 weeks. Beef tongues may be cured in with it but should be taken out at the end of 1 month, dried, wrapped in paper, bagged, and hung in a cool, dark place.

HOW *to* CARVE MEAT

One of the most important household duties is skillful carving. Meats and poultry are now so high in price that perhaps they will be more carefully considered than they once were. Whatever the price, neat and intelligent carving not only adds to the attractiveness of the food but also prevents waste.

Without good tools, acceptable carving is an impossibility, so every household should possess one, if not two, carving sets of standard make. If there is but one set, let it be of medium size. A better plan is to have one fairly large set and another of smaller size, known as game or steak carvers. The knives should be kept in the best condition, for a dull one is an abomination; therefore, it is no less important to be able to use the steel than it is to use the knife. Every once in a while the edges of the knives should be ground, for nothing dulls sharp edges so quickly as hot fat.

It helps greatly to have the meat put into shape before cooking by means of twine and skewers; a set of the latter, in steel, should be found in every kitchen. A large platter is also essential, for the carver must not be cramped for room.

While beef is probably served more than any other meat, the cutting varies in different localities, and what is known as a sirloin steak in one town, may, 100 miles [161 km] away, be called a porterhouse. Whatever the cut or the name, there should be one invariable rule: the meat should always be cut across the grain, not with it. With the right kind of knife the meat can be neatly cut away from the bones and should, in case of a roast, be crosscut in thin slices; if a steak, it should be cut in neat strips. Since people's tastes vary, be sure to ask their preference when serving.

A leg of mutton or lamb should be dished with the unsightly end of the bone covered with parsley or watercress. Stick the carving fork firmly into the small part of the leg and cut wedge-shaped pieces. Remove the first few slices and continue to cut in this direction until you reach the larger bone; then cut and loosen the slices afterward. When a shoulder of lamb or mutton is roasted whole, carve in wedge-shaped slices.

A slice of ham seems a homely thing to manage, but countless mistakes are made in serving it. It is often cut lengthwise, then cut into smaller pieces, and the choicer side served to those who like fat, and the other side to those who do not. The proper way is to cut the ham in vertical strips, so that each person receives his share of the part that is choice and that which is less so.

Bountiful serving has, of necessity, become a thing of the past. Along with skill in carving there should be a knowledge of the tastes and capacity of those who are being served. Small portions with a second helping are preferable to heaped-up plates, the contents of which are not consumed. While the daughters of the home are being trained in housewifely arts, the sons should have their turns at the carving and serving. Their efforts in this direction deserve recognition, for they have their place in the well-managed home.

PART IV

AROUND *the* HOUSE

MAKING A WINDOW GARDEN

What follows is an excellent example of how a window garden can turn a dreary setting into one of cheer.

It is late afternoon in December. The dark gray clouds, the cold drizzling rain, and the wind sighing through the naked tree-tops oppress me. But as I step from gloom without into my living room, sadness changes to joy—I have entered a garden, a small garden, to be sure, but a real garden, nevertheless. Before me is a large bay window framed by a gorgeous mass of color and foliage. Begonias, geraniums, coleus, heliotropes, fuchsias, sweet alyssum, scarlet sage, impatiens, and primroses shed their gladness and beauty about them, while the tender ferns, palms, and climbing ivy cheer me.

START LATE IN OCTOBER

I started my winter garden about the end of October, before Jack Frost had become a constant caller. The first thing done was to accustom my plants to their new environment. Therefore, I selected the choicest and most vigorous plants from the beds and borders outside the house and put them into pots and boxes, according to their sizes, being careful to place some coal ashes to the bottom of each to prevent the soil from becoming sour.

The soil used for this work consisted of a mixture of one part well-rotted manure to three parts of the best garden soil I could find. Selecting a sheltered but sunny corner of the front porch, I kept the plants there about 10 days, during which time they received the benefit of full sunlight and careful watering. Here they made a good root growth and were better able to withstand the sudden change from the garden to the house.

The bay window of our living room opens to the south, southwest, and southeast. I have noticed that the latter exposure suits the plants best, as they receive the brightest sunshine from ten in the morning to one in the afternoon. When I measured the width of the sills I found them a scant 6 inches [15 cm]. I widened them to 18 inches [46 cm], using chestnut lumber (which stands dampness well), planed and painted two coats of olive green. This extra width of sill permits the growing of several rows of plants, the small plants nearest the glass.

NO EXPENSE TO SPEAK OF

Instead of buying expensive saucers and pans to retain the excess water, I set the pots and boxes upon a 3-inch [7.5-cm] layer of sphagnum (florist's) moss, packing the moss well about them until only the rims were visible. The moss not only absorbs all the drainage water, but prevents the pots from drying out quickly. Another thing in favor of the moss is its attractive appearance. A very natural effect is obtained by tacking strips of bark on the front and back edges of the sill and allowing ivy or periwinkle to hang over.

In arranging my plants for the most artistic effect, I trained the climbing vines, such as smilax and morning glory, along the sides of each window sash, in order to hide the woodwork as much as possible. These windows are opened only upon the mildest days. The room is aired by opening the windows and doors of adjacent rooms, while the plants are covered with muslin or newspapers to protect them from drafts.

The temperature of the living room varies between 60 degrees F [15.5 degrees C] and 75 degrees F [24 degrees C], although I aim to keep it as near 65 degrees F [18 degrees C] as possible. A pan of water placed on top of the stove fills the air with enough moisture to prevent too rapid drying of the leaves. There is a great danger of overwatering plants before the roots are able to withstand it. Only sufficient water should be given to prevent the foliage from wilting. Later on, in late winter, when the sun becomes brighter and the pots are filled with young roots, water may be given freely.

THE NASTY ROACH

The ordinary roach or cockroach comes of an ancient (although not distinguished!) lineage that dates back further than the "family tree" of the proudest human family. Fossil remains prove that this insect existed way back in the Carboniferous age.

About five thousand different species of the roach family are believed to exist in different parts of the world, most of which live outdoors and subsist on vegetation. Only a very few species are engaged in making trouble in our homes.

HABITS

Roaches are lovers of the dark; it is then they roam around pantry and rooms in search of mischief and food. At the approach of light they scurry away, like a pack of cowardly thieves. Any kind of food tastes good to them, whether it be shoe leather, apple dumpling, or book covers. Toothache, loss of appetite, or the common cold are, we believe, unknown to the roach family. But the roach has one good quality. (Alas, only one.) It is this: he has a fondness for bedbugs; by eating them, he often does homes a real favor.

REMEDIES

Now we come to "remedies," and here the trouble begins. Nothing short of eternal vigilance will rid a house of these pests once they gain a foothold.

Unfortunately the roach seems to be endowed with remarkable intelligence when it comes to poisoned foods. Arsenic, no matter how disguised, he refuses (with thanks) nearly every time. However, it is said that a preparation of sweetened flour paste containing phosphorus will often fool him.

Another remedy often used is fresh pyrethrum powder. This, when liberally dusted on shelves, etc., usually affords temporary or partial relief. A better use of this powder, however, is to burn a quantity of it in an infested room and then tightly close the apartment for 10 hours. Carbon bisulphide is sometimes used in this way also, but its vapor is more dangerous to have in a house and can cause skin irritation and burns. Some of the prepared "roach powders" that are on the market are also effective when perfectly fresh.

Trapping the insects is another remedy. Roach traps may easily be made at home as follows: Take any deep vessel or jar and place it where the roaches congregate. Fill it partly full of sweetened liquid paste. Then take several thin, narrow pieces of wood, bend each one into an inverted *V*, and hang them on the jar—one end almost in the liquid, the other on the shelf or floor. The idea is to make several "gangplanks" up which the roaches can crawl, with a steeper gangway inside, down which they will slide into the liquid—never to return.

PAPER WALLS *for* WARMTH

Almost every home has at hand the means of making farm buildings warm. Tack on coat after coat of old newspapers. Then board, shingle, or clapboard directly over them. Air simply cannot pass through successive layers of paper. If desired, red building paper can be put over the newspapers, so if water gets through the boarding, it will be carried down to the ground by the resin-sized paper.

PAPERHANGING MADE EASY

In certain sections of the United States, "papering boards" were at one time considered as necessary a part of the household equipment as quilting frames. Both fell somewhat into disuse, but the pendulum swings both ways, and many householders are now finding it expedient to do their own wallpapering.

Fig. 1

The work requires few tools and appliances, the first requisite being two boards 7 to 8 feet [2 to 2.5 m] long and 10 to 12 inches [25 to 30.5 cm] wide. Place them side by side with the ends resting on a table about 3 feet [1 m] high. This will make an excellent paste table. The other articles needed are a paste brush, paperhanger's smoothing brush, a seam roller, and a pair of shears or a straightedge and trimming knife.

Paper the ceiling first. Measure the length of your room, unroll the ceiling paper, pattern up, cut, and match enough for the whole room, being sure to leave the strips 3 to 4 inches [7.5 to 10 cm] longer than the length required, so that when hung, 2 inches [5 cm] will come down on the walls at each end.

Then turn the paper, pattern down, and it is ready to paste. Beginning at the left, apply the paste evenly. After you have pasted half the length, fold it over, being careful to see that the sides are perfectly even. Then paste the other half and fold it toward the center, just as you did the first half. Now you have a full strip pasted and ready to trim.

Fig. 2

Draw the paper toward you about 3 inches [7.5 cm] from the edge of the table and proceed to trim, commencing at the right. The great advantages of trimming after pasting are that the paste is more evenly distributed at the edges and it takes less trim. As a guide to hanging the first strip of ceiling paper properly, drive a nail in at each end of the ceiling 16 inches [40.5 cm] from the side walls, chalk a piece of cord with charcoal or chalk, and tie to the nails, being sure that it is drawn tight, then take hold of the center of the cord, and pull it down and let go; the cord will strike the ceiling and leave a chalk line. Take the first strip you have pasted and trimmed, unfold the end to your right and let the other end, which is still folded, hang over a roll of wallpaper that you hold in your left hand. Guide the paper along the chalk line, at the same time pressing it flat on the ceiling with your hand as you move along, and smoothing it out with a smoothing brush. When half the strip is on, unfold the other half and continue to the end.

Fig. 3

When the ceiling is finished, cut the side wallpaper, matching it and leaving 4 to 5 inches [10 to 12 cm] to allow for any waste in matching. The upper ends of the strips will be uneven, but these will be covered by the border. The lower ends, which will stop at the baseboard, are either trimmed off with a base trimmer or a pair of shears. When using shears, paste the paper down in place close to the top of the baseboard. Use the back of the shears for marking by running them over the top of the paper where plaster and baseboard meet, lift the paper a little, and trim along where you have marked and smooth down again with a smoothing brush.

When hanging the side wallpaper, commence at any door, as corners are often irregular. This will assist you in hanging the first strip straight; then continue around the room until it is finished. The short pieces can be used over the doors and under the windows.

Fig. 4

The border should be hung last, and may be cut in five or six pieces, making it easier to hang. Paste, fold, and trim the border just as you have done the paper used for side walls and ceiling. A substitute for the usual wheat paste may be made with one part by weight of dried glue in ten parts by weight of water, melted in a glue pot surrounded by boiling water. To this should be slowly added four parts of laundry starch stirred up with ten parts of warm water. This produces a perfectly smooth paste, the consistency of which can be varied by changing the proportion of water used. If the paste is to be kept for any length of time, some preservative, such as oil of cloves, oil of wintergreen, or oil of sassafras can be added.

Fig. 5

If wallpaper is marred or broken in places, it may be repaired by cutting from a remnant of the same paper figures or groups of figures corresponding to those that need repairing. The outline of the design should be carefully followed in cutting, and the whole should be matched and pasted exactly. If the new paper used to repair is hung in the sunlight and faded a bit, the mended places will escape notice. The colored crayons used by school children can, if they match, be rubbed over small breaks. The paints that come in toy paint boxes will serve the same purpose.

ONE SOLUTION
of the
HOUSEKEEPER PROBLEM

Through accident and illness the father's health failed, so there were two reasons for doing without a housekeeper: limited income and lack of girls willing to serve in any household except that of their parents.

The mother had been taught in childhood many ways of economy, and imparted this knowledge to her children, little by little, as occasions were presented.

DOING WITHOUT

There were two girls and five boys, but the older daughter taught school for a year, helping with fertile brain and deft hands during vacations; then she married. The daughter left at home assumed one duty and then another, finally including the laundry. Then the boys, who were accustomed to cutting, piling, and bringing in the wood, mowing the lawn, etc., had to wash dishes. Sometimes the mother was ill; then the big sister gave orders, the boys (all younger than herself) obeying with more or less readiness.

It became the duty of one boy, of twelve, to wipe the stairs and banisters and oil the floor. The lad who, during his first years of voluntary locomotion, had been always at his mother's elbow to upset things when she was "mixing," made delicious cakes and doughnuts; and once, in a week's experience of "home without a mother" said: "I'm sick of father's bread. I can make bread!"

(Ah! What stupendous power in the determination expressed in the four letters, "I can!") He made bread—good bread—for years, becoming an expert. It didn't spoil his muscles for throwing a ball, nor his brain for study.

By being much with mother and sister, they all soon knew how to make coffee, fry potatoes, cook meat, and make pancakes, and one fourteen-year-old chap made cake fillings.

They could all sew on buttons, sew up rips, and press their clothes, although mother and sister attended to the more difficult jobs of mending.

The boys did errands, of course, saved journeys upstairs and down cellar, whipped rugs, washed windows and floors, and even tinkered up things sometimes. A boy likes a little responsibility; it helps his self-respect.

HOMEMADE DISHMOP

Take any smooth, even stick about 1 foot [30.5 cm] in length. Notch it 1 inch [2.5 cm] from the bottom. Take several pieces of good cotton cloth, 6 or 7 inches [15 or 17 cm] square. Place the end of the stick in the center of the cloth, then bring the cloth up about the stick and tie in the notch; next turn the cloth down and tie about the bottom of the stick with stout twine. This makes a serviceable dishmop at practically no expense.

THE BABY
and
HIS BELONGINGS

Certain things are necessary for the comfort and development of a baby, among which are quiet and clean surroundings, fresh air, proper food, and clothing. Other necessities are a separate bed, a small bathtub, and a basket for baby's toilet accessories. The baby should be trained to sleep by himself from the hour of his birth. Not only will he rest better, but so will the mother, and the chance to form bad habits is eliminated.

SUBSTITUTE BEDS

A very cozy nest can be made in a clothes basket or, in a pinch, a bureau drawer could be used. The basket bed has many advantages. It is easily moved about, may be placed upon two chairs or upon a low stand, and will accommodate the baby for several months. The bed in the clothes basket has a bag filled with finely shredded corn husks in the bottom. Over this is placed a folded quilt, and over this may be laid a thick pad of cotton or several layers of silence cloth. This is covered with rubber sheeting, then the small sheets, soft blankets and the white spread, which is made of 1 yard [1 m] of crinkled cotton crepe.

SELECTING A CRIB

In choosing a crib, select one having the bars close together, which is absolutely plain (no fancy parts to come loose). Brass rods across the top are objectionable, for teething babies invariably cool their aching gums upon the crossbars of their cribs and an enameled rod is preferable for this purpose.

SLEEPING BAG

For a winter baby, the sleeping bag of outing flannel will keep the little hands and feet warm. The bag is simply made of a piece of flannel, 22 inches [58 cm] wide and 6 feet [1.8 m] long. Fold the flannel crosswise, cut a shallow piece out for the neck, cut down the front, and face it. Bind the neck, put in a drawstring, and add buttons and buttonholes down the front. The flannel is then doubled, the edges seamed together, and the seams catch-stitched. The bottom edge is hemmed for a casing and finished with tape and drawstrings. Such a bag allows freedom of movement without exposure.

THE BATHTUB

Quite as important as the baby's bed is his bathtub. For the first few days, baby may be sponged upon the parent's lap, but after that he should be put in the tub for his daily bath, so that he may become accustomed to it before he is old enough to fear it. For emergency use, a footbath will answer and, when that is outgrown, a small washtub of galvanized iron or papier-mâché might be used.

The baby's parent will like to wear the apron, requiring 76½ inches [194 cm] of outing flannel. The ends are hemmed and brier-stitched with white mercerized thread. The upper part is slightly shorter than the under, and a casing is stitched across the top. Cotton tape 1 inch [2.5 cm] wide is run through this for strings and fastened with a few stitches right at the center to prevent its slipping out. In lifting the baby out of his bath, he is placed between two parts of the apron and kept there while he is being dried.

BASKET FOR BABY'S THINGS

A plain basket large enough to hold the things necessary for baby's toilet should be provided. Lacking this, a strong oblong pasteboard box could be neatly covered with pink or blue cambric and, if you choose, white swiss or lawn. The basket should contain talcum powder, a tube of white vaseline, a roll of absorbent cotton in a tin box, wooden toothpicks in a small bottle (for swabs), several sizes of safety pins, soft towels, and washcloths. Soap should be kept in a celluloid soapbox.

EARLY TRAINING

Along with the tub bath and separate bed, baby should be accustomed to using a toilet. As soon as he is old enough to sit up, he should be regularly placed upon the "baby's throne" and every effort made to train him to regular habits. A good makeshift baby toilet can be fashioned out of a box by cutting a hole in it for the seat and building up a back and sides with smoothly planed boards. It may be padded with an old comforter, folded to fit, and a cushion covered with white oilcloth for a seat. A hole through which you can slip a strap or a strong piece of webbing should be bored in each side of the chair, so that the baby can be securely held there. By placing this chair over a vessel, it serves its purpose admirably, and its use will save the mother much labor.

Directions for the preparation of baby's food usually call for a double boiler, which is not always available. To improvise one, place a wire dishcloth in a dishpan or other large pan. Upon this, stand the vessel containing the food that is to be cooked, and surround it with plenty of water. The wire cloth between the pans keeps them from sticking together and the contents of the inner vessel are never scorched.

SONNY'S BATH

"Come in!" cheerfully called out the young neighbor, in answer
to the old-fashioned mother's knock. "You are just in time to see
Sonny have his bath."

"Perhaps I better not," the caller answered, at the same time
closing the door behind her, "won't he make an awful fuss?"

"Not Sonny," the little mother replied. "He just loves his
bath. Why, it's our frolic time. Eh, little man?"

In answer the baby waved his chubby arms, kicked, smiled,
and emitted a series of sweet, cooing sounds.

The visitor was astounded. "He'll cry before you are through
with him, I bet. You're the first mother I ever heard of who spoke
of a baby's bath as a frolic time! My babies always screamed from
the moment I took them up to bathe them until I had finished.
It was my day's hardest task, and I was always thankful when it
was over."

"I don't think he'll cry," was the mother's only answer. "See
how good he is while I wash his eyes, nose, and mouth."

The older woman watched in amazement. While they were
talking the young mother had put a teaspoonful of boric acid
into a cup of warm water. Now she pulled tiny bits from a roll
of absorbent cotton. One of these she dipped in the water and
carefully squeezed a single drop from it into each eye, quickly
wiping the eye with a dry bit of the cotton. The baby gurgled
and laughed. Keeping the baby's attention all the time, with deft
fingers she squeezed a bit of white vaseline on two more swabs
of cotton, twisted them firmly, then carefully cleansed each
nostril, using a separate twist for each. Again, he laughed.

It took but a moment to wash the rosebud mouth. Baby's
mother wound a piece of the cotton around the end of her little

finger, dipped it in the boric acid solution, and while baby bit at her soft finger, washed tongue, gums, and lining of the mouth.

"Well, I never!" the caller said. "I never went through all that for my babies. It's lots of work, isn't it?"

"Yes, it does take extra time, but it's worth it. Baby has never had sore eyes or mouth, and his little nose is so clear he can always breathe through it."

"I wish I'd know that when I had babies to take care of. They always had sore mouths, and sometimes red, inflamed eyes. But we thought that was as common with the babies as cutting teeth. As for the nose, when I saw it was dirty, I cleaned it with a small hairpin. The youngsters always fought against it. I suppose it did hurt."

The young mother shuddered at the very thought. "Ah, now he'll cry!" the caller exclaimed, "when he gets soap in his eyes!"

But no soap was used on his face. It was carefully washed with clear water and patted dry.

Until then the baby had been fully dressed. Now his mother removed his clothes—kimono, flannel petticoat, shirt, binder, and diapers. "I always take off his nighty, which is apt to be damp, the first thing in the morning, and put on a warm flannel kimono. He is never fully dressed until after his bath always at half past nine."

The old-fashioned mother thought of her babies, who had lain and fussed in their nightclothes when she was ready to bathe them. Perhaps, she pondered, that may have been one reason why they were so cross during the bath.

Sopping a wet cloth with castile soap, she washed first the back and then the front of the baby and, while the caller stared with wide-open eyes, lifted him gently into a tub of water. With the fingers of her left hand spread to support the tiny head and shoulders, she rapidly rinsed off all the soap with a wet sponge, and in the twinkling of an eye had the baby again in her lap, face downward in the large soft towel she had pinned to her left side, and almost enveloped by the free end of the towel that the mother had thrown over his body.

The visitor gasped. It had all been done so quickly, yet so thoroughly, without a murmur or dissent. Instead, peeping out turtle-fashion from the towel were two bright eyes, gazing at the visitor's red shawl, while their owner contentedly sucked a moist pink arm. A gentle patting with the bath towel, a careful drying of all the creases, a brisk rubbing of the scalp, and then a slight dusting of powder in chafeable spots—and Sonny was ready to be dressed.

Once more the older woman claimed, "Here's where he'll cry!"

But again she was wrong. There seemed to be no bungling, hard-to-put on clothes. Instead of the tight bellyband that she had always dreaded to sew on, this mother slipped over the youngster's feet a knit band with shoulder straps. The shirt was doubled-breasted and fastened with one small safety pin. The petticoats were slipped into the simple little dress and as one garment were drawn over the feet. Baby was turned face down-ward, and the three garments were buttoned without further disturbing the wearer. He actually enjoyed it.

When at length the little mother brushed back his silky down of hair, and, after wiping her nipple with a piece of cotton saturated with the boric acid solution, placed him at her breast, she turned to the visitor with a happy smile. "Do you wonder I enjoy this hour?" she asked. "Sonny is always like this at bath time. He is never tired or hungry at half past nine; I have every-thing ready so I won't have to make him wait, half dressed, while I find some necessary thing; the water is always the same temperature—98 degrees F [37 degrees C]—so he receives no shock when I place him in the tub; and most of all, he feels how much I enjoy it, and so has confidence in me. Now he'll nurse and go to sleep."

"It's well-nigh wonderful," the old-fashioned mother replied. "I'd never have believed it could be done if I hadn't seen you do it. Bathed a baby—put it in a tub of water, even—and it laughed and cooed and kicked its legs and waved its arms in glee all the time!"

MADE-OVER FARMHOUSE

The farmer's wife rolled up her ball of olive-drab yarn, and laid it carefully aside. "Mary and the children are coming to live with us," she said. "Of course, the house is entirely too small for such a large family; so my husband wrote you to come here and tell us what to do."

"H'm, yes; let's see." The architect sat down at a table and sketched rapidly for some minutes. "Here you are," he said, at length. "This first drawing is the exterior of the house as it now stands; the second drawing is what it will look like if you follow my suggestions.

"Very simple; we just run the old rooflines right on down to the first story, front and back. As for your floor plan, your current living room will never be big enough for your enlarged family, so I've added a den at the rear and cut a big archway to connect it with the living room. I've built on a bedroom, too, as you see."

Fig. 1

"What's that little place marked 'lav,'" asked the farmer's wife.

"A lavatory; you'll find it mighty convenient. For the second story, I'll fix you up a splendid pair of new closets under the slant of the new rear roof. The bathroom is also worked in under this new roof."

"But I wanted another bedroom up here."

"Exactly; see that big new sleeping porch! In summertime it will make the best sort of bedroom for your grandchildren; in winter you'll enclose it with glass, and go on using it, and it won't be so very expensive, for I've left the old house practically untouched, and merely added things onto it. Changing things would cost money."

Fig. 2

PLANNING *the* PORCH

"... Just exactly like *that!*" finished the farmer's wife, flipping the photograph down on the drawing board.

"But, my dear lady ..." protested the architect.

"No, that's the porch I want and you can't talk me out of it!"

Silence, then a little laugh from the farmer's wife. " Come now—*why* don't you like that porch?"

The architect opened a drawer and lifted from it a huge handful of photographs; carefully he spread these out before him. "Why? Because it falls into one of the four classes of badly planned porches. Now here are seventy pictures of seventy farmhouses—no, I didn't design them!" He laughed as he hastily sketched several plans on the back of a photograph. "Here you are," he said, as he showed her the sketch. "Only a portion of the porch can get the breeze; half of the space is pocketed by the angle of the house. Sixteen of my farm pictures show just that defect.

"But I *like* a porch that turns around the side of the house!" The farmer's wife smiled a bit defiantly. "You can always get a good breeze and you can keep out of the sun, too!"

"Quite true, but I'll explain presently how you can get both these advantages in another way without the excess cost of so much porch. Besides, a round-the-corner porch usually covers all the windows of one or more rooms; very dark and gloomy, that! And here are twelve houses with porches all across the front; that would be well enough if only these weren't so narrow. Remember, a porch is really an outdoor living room; especially so on the farm. Now, the *very narrowest* room is at least 10 feet [3 m] wide;

yet we often see porches of 7 feet [2 m], or even less. You can't form a family group; you string yourselves out in a row, minstrel-fashion, and talk back and forth as best you can. Then, there are ten more with merely a cramped little entrance porch, and finally seven porches that really are porches. Seventy in all; which means that only 10 percent are really good!"

The farmer's wife gasped. "But—I don't quite understand; what makes those seven so good?"

"Well, they average 12 feet [3.7 m] deep and 20 feet [6.1 m] long. That makes a very cool, comfortable outdoor room, indeed."

"But supposing the breeze should change?" was her prompt question.

"Well, let's suppose the house faces south; then you get all but the north wind, and you don't want that!"

"But if it doesn't face south?" insisted the farmer's wife.

"Then we will put just a little entrance porch at the front and tack our main living porch to the south end of the house. See?"

"Yes, I see!" admitted the farmer's wife.

A STANDARD FARMHOUSE

"Blueprints from the *Farm Journal* office," said the draftsman, laying a bulky bundle on my drawingboard. Blueprints? That was strange; what did it mean? I opened the letter from the editor, and read a few lines.

" . . . So I'm sending you these standard farmhouse plans, just received from the University of Missouri; if you think best, you might let Our Folks see them."

Then I understood! And very interesting the floor plans were, too; simple, compact, inexpensive. But some few of the arrangements seemed as if they might be improved a trifle and the outside of the house was a bit too bare and barnlike.

So, as architects will, I have used my blue pencil on this design; cutting out something here, marking in something there, and so on. I think I have improved it; but quite possibly Professor Fenton (who first designed it) may think just the opposite!

On the first floor, the closet beneath the stairway gives a passage from kitchen to bedroom; any housekeeper will grasp the comfort of this.

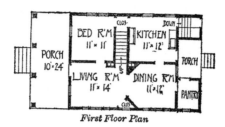

First Floor Plan

The cellarway goes down from the kitchen, and has an out-side door at the ground level; you get the idea? In the second story the balcony is meant as a place for airing bedding, moving furniture while housecleaning, and so on; but it's also a fine place to sleep on hot summer nights. (I know, for I have one on my own farmhouse!)

The storage space at the rear is rather bigger than I've shown; it really runs out under the eaves and gives a lot of room to pack away all sorts of useless stuff.

The chimneys run up separately in the first and second stories, but rack over and join before passing through the roof. If there is a cellar and a heating plant, one of these chimneys may be cut out; but I know from experience that there are some places where the water lies so near the surface that a cellar is simply out of the question.

I have imagined that the house is built of frame, covered out-side with cement stucco on metal lath; but of course, one might use almost any other material.

Just one final word: let the front porch face south. That brings the cheery morning sunlight into the dining room, shelters the living room from the cold northwest winds, but lets the summer breezes blow across the porch and through the front bedrooms.

THE NEW FARM HELPER– ELECTRICITY

A man who was called upon recently to speed up the production in a big factory tells that in one instance, by the introduction of a little inexpensive machinery and the rearrangement of the working force, three men were able to do the same work that eight men had been doing before and do it better.

On a great many farms, a force—new as a factor in farming but wonderfully practical and efficient—is being introduced to help solve the labor problem and to help speed up production. This force is electricity, and the introduction of small individual electric power and light plants makes it possible to use electric current anywhere.

This is an important matter for the farmer. He knows better than anyone else about the difficulties that go with an attempt to secure manpower for farm labor, and to keep it, once it is hired.

Electric power on the farm helps out in more than one way. It does many tasks that otherwise must be done by hand, freeing the hands for more profitable and often more congenial labor, and it furnishes improved conditions of living that appeal to the owner's family, to the boys and girls and to the hired help, and make it easier for them to be satisfied with conditions on the farm.

DRUDGERY TAKEN OUT OF CHORES

Electricity is influencing the farmer himself in a vital way. Life on the farm with modern conveniences in the house, and with the drudgery taken out of the chores, does not become distasteful as he grows older. He is content to stay out of the "retired farmer" class, and to live out his life in the farm home where he can keep his mind and body healthfully occupied and where the benefit of his advice and experience can still be had by the younger workers.

The benefits that electricity bring to the farm home are tremendous. Electric light about the barn is a time-saver. The chores can be done after dark with ease, and dispatch when securing the light needed is just a matter of snapping switches off at convenient points instead of carrying a lantern. With electric light in the farm buildings, it is easy to use all the daylight hours in the field and do the chores after dark.

SAVE ENERGY ON CHORES AROUND THE HOUSE

Then electric power can be used in various ways to lessen labor of many jobs, as in the matter of running a churn or cream separator, a grinder, feed chopper, grindstone, fanning mill, and the like. It is proving quite a boon in the operation of pressure pumps for pumping water for the stock, for sprinkling the yard or garden, and for household use, doing away with a lot of hand pumping. It is running milking machines, cutting the time of milking in half and doing away with the mighty unpleasant task of milking by hand.

It is running washing machines and vacuum cleaners, saving time and lightening the load of household chores; likewise electric fans—destroying the ill effect of sultry weather and keeping everybody fit for more and better work.

An electric power and light plant will run one of the small motors that will operate ordinary light machinery, for 2 cents an hour if the fuel is gasoline; if kerosene is used, the hourly cost will be about half that. In many farm homes where electricity is used, a time saving of 20 to 25 hours a week is frequently reported. This is a considerable item, especially in seasons when labor is hard to get and when time is precious.

"MERELY BY TURNING A SWITCH"

It is easy to operate your own electric power plant. There are plants on the market for which the only attention required under ordinary conditions is to start the engine once or twice a week, see that fuel is provided for the engine, and add a little distilled water to the storage battery occasionally.

There is a strong appeal in the possibility of electricity to the farmer and to those who labor with him. They have been accustomed to accomplish things by the outlay of actual strength, by toil that tries the muscles and oftentimes oppresses the spirit. To be able to accomplish the tasks merely by turning a switch and then watching the mysterious electrical force go ahead and do that task silently, tirelessly, efficiently, and well has a fascination for the farmer just the same as it has for those in any other line of business who realize that there is a benefit to letting a machine do all the work that does not have to be done by hand.

So we are going to witness—are witnessing—the introduction of this modern force on the farm at a rate that foretells great and helpful changes in labor conditions there, with a corresponding benefit to the producer.

ECONOMY
in HANDLING
the WASHER

While the washing machine has come to be appreciated as one of a homeowner's best friends, it deserves better care and attention than it receives on the average farm. On many farms a new washer is bought every few seasons, but with proper management, economy could be brought into practice along this line, as washers are very strongly made and with good care will last several seasons.

A good coat of paint would make the old washer look like a new one and guard against rust and decay. Keeping a bucket of water in the tub of the washer at all times will prevent bulging and warping of the bottom and shrinking of the staves.

A LIGHT LOAD IS BEST

About a half tub of water and a small amount of clothes require the minimum turning of the machine and is a light strain on its working parts. When the washer is jammed full of clothes, it greatly increases the labor of the person operating the machine, while the clothes will not be clean so well as a few.

Frequent adjustment of all the bolts, screws, and other parts of the washing machine should be made, as it turns much harder with these things loose and is a damaging strain on the whole machine.

OIL PERSISTENTLY

Oiling the washing machine every time one uses it makes wash-day less of a drudgery and prevents wear and tear of the machine, as well as adding to its length of use. Run kerosene through the gearing once a week to cut loose and remove refuse grease, oil dirt, etc. Then give another thorough oiling with good separator oil after the cleansing. Use only first-grade oil.

INDOORS, PLEASE

Above all, keep the washing machine in out of the weather. Either have a rainproof building in which to do the washing, keeping the machine there at all times, or store it in a dry place after each washing. Left outdoors, the action of the sun, wind, and rain on the washer will soon deteriorate it in value till it is practically worthless. Good care in this respect, together with proper handling and oiling, as directed, will add much to its life, effectiveness, and easy-running qualities.

DISHWASHING MADE EASY

That same old hateful task of washing dishes—how we loathe the time and effort that would be so much better placed on something else!

But dishwashing is being touched by the hand of science; a few years hence dishwashing drudgery will have disappeared forever. Even today, in the most convenient kitchen, the task may be made easier by applying a few simple methods of work.

Analyzing the process of dishwashing from the time we remove the soiled dishes from the table to the time we put them away on their respective shelves, we see that dishwashing is not one single task but a group of small tasks. We see that it is not all "dishwashing," but that it is composed of these four parts: 1. Scraping and stacking dishes. 2. Washing them. 3. Drying them. 4. Putting them away.

The reason the work is hard may be due to fault or delay in any one of the four parts, and we have to find out in which part the delay occurs, and how it can be remedied, before we can improve the whole. Perhaps we do not thoroughly and carefully stack and scrape. Few use the little scraper of wood and rubber, which looks like a paddle (which is better than a knife or fork), to scrape the dishes clean, so that the actual washing is not bothered by fragments, grease, etc. Or we can use tissue paper to remove grease and thus make the work easier. Stacking is important, all similar sizes going together, all dishes stacked to the right and drained to the left.

Nothing does more to make dishwashing a drudgery than to have the sink, or table surface on which dishes are washed, too low. The height should be so arranged that the worker can stand at it with comfort, or, preferably, the worker can sit down to this, as well as any other household task of a similar kind. The writer always sits down to wash dishes, iron, prepare pastry, and peel vegetables.

The first secret of washing dishes is to have the water hot and soapy, the suds being made by a soap shaker and not by floating a cake in the water. A good soap powder is best for pots, kettles, and pans.

Now, if we have a drainer, we can cut down the time of the whole process more than a third, because the time consumed in wiping is greater than the time consumed in washing. Therefore, with a drainer, we reduce the drying time, because if dishes are arranged in a rack and have scalding water poured over them, they will dry themselves. It is neither necessary nor sanitary to dry each piece with a dishtowel of doubtful cleanliness, and by the time the pots and small utensils are finished, the dishes will be completely dried and ready to be put away without further handling.

Much time is consumed in carrying the trays of washed and dried dishes to some distant place. Sometimes innumerable trips are required to carry all the dishes used at a meal to the pantry or shelves. Why not place the shelves for the pots and pans, and the kitchen dishes, adjacent to the sink, at the left? This has been done in several homes with the result that the dishes can be put away, when dried, without walking a step. Thus many minutes can be saved to be spent on something else, outside the kitchen. A device that can be helpfully employed in the dishwashing process is a tray on wheels, which can be pushed or wheeled wherever desired. All the dishes from a meal can be stacked on these lower and upper trays and wheeled at once to the right of the sink. If the dish closet is some distance from the sink, the same tray can be used to receive the clean dishes and wheel them to wherever they are to be put away.

A new dishpan that can now be bought is rectangular in shape, more exactly to fit the sink. It is supported on legs, which are fitted with rubber tips to prevent marring the surface of the sink. In addition, it has a small mesh wire. Thus, instead of lifting up and tipping the pan to empty the water, the stopper can be removed, the drawer pulled out, and the water emptied, and the pan drained at the same time. This is a great improvement over the old round pan. With a convenient arrangement of sink, closet, and shelves; proper tools; and a good supply of hot water; time, strength, soap, and towels can be conserved, all of which are worth saving.

CANNING MADE EASY *by the* COLD PACK METHOD

Here is the standard method for canning fruits and vegetables, as advised by the United States Department of Agriculture. Wash the jars; wash the rubbers and test them for quality. Set the empty jars and rubbers into a pan of water to heat and keep hot.

Use only fresh, sound vegetables. Wash, then place in colander; blanch by setting in a tightly covered vessel of boiling water or steam for 5 to 10 minutes for beans, 1½ minutes for tomatoes, 5 minutes for sweet corn and beets.

Remove from the boiling water or steam and plunge quickly into cold, clean water for a moment only. Remove and pack immediately into hot jars; add hot water and seasoning, place rubbers and tops of jars in position, screw them down tight, and then give a quarter of a turn back. Place jars on a false bottom of lath or wire placed in the washboiler, submerge them 2 inches [5 cm].

Put the cover on the washboiler and let the water boil hard for 2 hours for beans, 22 minutes for tomatoes, 3 hours for sweet corn, and 1½ hours for beets. Start counting time when the water begins to boil. Remove the jars, tighten the covers, invert to cool, and examine them for leaks. If leaks are found, change rubbers and boil again for 10 minutes.

Fruit that is to be canned by this method should be sound and fresh. Wash it carefully and remove any parts that are decayed. Place all fruit, except berries, in a square of cheesecloth or wire basket and dip into boiling water for 30 seconds for peaches, and 1½ minutes for apples and pears. Plunge for a moment into cold water. Skin the fruit if necessary and leave whole or cut as preferred.

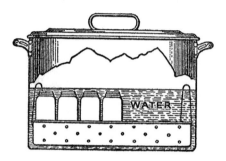

Pack the fruit in hot jars, fill the jars with hot syrup or boiling water, and proceed according to directions for vegetables, allowing the water to boil 16 minutes for peaches and 20 minutes for apples and pears. Make syrup according to formula 1, which requires 12 cups [2.4 kg] sugar and 8 cups [2 L] water. Boil until the sugar is dissolved, skim off impurities, and keep it hot; or use formula 2, which is not so sweet. This takes 8 cups [1.6 kg] sugar to 12 cups [2.8 L] water.

Imperfect jars, caps, or rubbers are responsible for the greatest percentage of spoilage. It is quite necessary to provide a good storage place that should be both cool and dark. If exposure to the light is unavoidable, wrap the jars in brown paper, or store them in the boxes in which they were bought.

Commercial canning outfits will in every case expedite the work and should be found in every home having a large quantity of food to prepare, or in community kitchens. The get-together spirit is necessary for a movement of this sort. It has been demonstrated that community kitchens produce the maximum of food conservation and offer the most economical means for canning, five people in a community kitchen doing the work of forty to fifty families.

INDOOR GAMES

CLOTHESPIN GAME

In this game the only materials necessary are two dozen clothespins and two small boxes. Two teams are chosen, each with the same number of players. The players sit on the floor in two straight, parallel lines facing each other. Each line represents a team. About 4 feet [1.2 m] should be allowed between the two lines, and the players should sit close together. A captain is chosen for each team and takes a position at the end of his line. Each player with his right hand takes hold of the left wrist of the player to his right. This hold must continue throughout the entire game.

A box containing twelve clothespins is placed beside each of the captains, who, at a given signal, picks out one clothespin and passes it to the player at his side. The pin is passed down the entire line until it reaches the player at the other end who places it on the floor by his side. Then another pin is started. Only *one* pin must be in action at a time. When all twelve pins are at the other end of the line, start them back in the same manner, and the team getting its dozen pins back in the box first wins the game. If any player releases his hold, it is a foul and the pin must be started again.

THE STANDING BROAD GRIN

Line up the contestants in a row all standing and facing the audience. At the word *go*, each must smile and hold the smile. The one who smiles the longest without moving his face wins.

THE CONTINUOUS GLUM

Line up contestants as in the previous game. At the word *go*, all look glum. The audience tries to make them smile by talking or making faces. They must not touch the contestants. The one who remains glum the longest is the winner.

FIFTY-YARD SLASH

For each heat have four strips of narrow paper, 1 inch [2.5 cm] or less in width, and at least 20 feet [6 m] long. Fasten one end of each strip securely; then the four contestants, each holding the free end of one of the strips, cut with a pair of scissors down the center without running off at the side. The one who reaches the fastened end first wins the heat. Any one running off loses.

CLEAR THE TABLE

The players sit in a circle and each one takes the name of an article used at the tea table, such as tea, sugar, cream, cake, bread, etc. The one called "tea" begins. He rises, turns around and around in his place, saying: "I turn tea; who turns sugar?" Sugar turns, saying: "I turn sugar; who turns milk?" And so on, till everyone in the circle is turning. They must continue turning till the leader claps his hands and calls out: "Clear the table," when all sit down in their chairs again.

GAME OF ARTISTS AND CRITICS

A good old game is "Artists and Critics." Furnish each player with a slip of paper and a pencil, and direct him to draw a picture of any sort he pleases at the head of the paper, then write its title at the bottom of the sheet. Usually the less it looks like what he calls it, the more fun. He must fold the paper up over the title so that no one can see it, then pass it to his neighbor, who writes what he thinks it is intended to represent, and folds his title under and passes it on around the room for each to add his criticism. When all of the slips are thus completed, someone collects them all and, first showing the sketch at the head, reads the various titles, ending with the artist's own.

GAME OF TELEGRAMS

Choose ten letters at random and put them up where all can see. Provide each guest with paper and pencil, and have each one write a telegram, keeping the letters in the order given, and letting each letter be the first letter in a word. Suppose the letters were A L P O C R G D E H. The result might be "All lazy preachers often come running gracefully down even hills" or "All ladies passing our car receive good dinners eaten hot." Nonsense, of course, but laughter provoking.

HEAVYWEIGHT THROW

Get a half-bushel basket and a cheap ball. A basket or tennis ball is best. Choose sides. The contestants stand 12 feet [3.5 m] from the basket. Each player endeavors to throw the ball into the basket. The side wins that succeeds in getting the ball to stay in the basket the most times in a given number of throws.

GAME OF BIOGRAPHY

Pass paper and pencils to all of the guests. Ask them to write their names at the top, and turn the paper down across the top so that the name does not show. Pass the papers around so that no one will know whose paper he has. Then ask the following questions, letting each member of the company write one answer on each paper by passing the sheet to his right after he has written and folded his answer out of sight: Where was the hero or heroine born—year and month? What were the child's first words? How was its youth spent? What was its first joy? First sorrow? Present occupation? Favorite fad, favorite food, favorite author, favorite book, favorite song, favorite statesman; politics? Where will he or she be ten years hence? Read the biographies aloud before the company.

INDOOR SPORTS

The hurdle race is for both boys and girls. Each person is given six needles and a spool of thread, and the one who threads them all wins the contest.

Next comes the standing high jump. Hang three doughnuts in a doorway, about 4 inches [10 cm] higher than the mouths of the contestants. Tie their hands behind them and see who first bites a doughnut.

For a drinking race, each player is given half a glassful of water and a spoon. The water must be consumed a spoon at a time, and the one who finishes first is the winner. If any is spilled, that contestant is barred out.

A bun race is great fun. A clothesline is stretched across the room, and from it are hung sugar buns at a height just reaching each player's mouth. The players stand in line with hands behind them, and at a given signal begin to eat the buns. The bobbing of the line makes this very difficult.

Last comes the rainy day race. Each contestant is given a shoebox containing a pair of overshoes and tied with string. A closed umbrella is also handed to each. When the starter counts to three, the boxes must be untied, the overshoes put on, and the umbrellas opened. The contestants then walk across the room as rapidly as possible to a set line, remove the overshoes, replace them in the boxes, tie the boxes, and close the umbrellas before they walk to their starting place. The one who arrives there first wins.

POTATO RACE

Arrange five chairs at each end of a long room, facing each other. Place six large potatoes on each chair. Five persons play at a time. The object is to carry each potato on a teaspoon from the chair at one end of the room to the chair at the other end, without dropping it.

The potato must not be touched except by the spoon. If it is dropped, it must be picked up by the spoon, carried back to the chair, and a new start made. The one who gets all his potatoes over first wins.

NUT GAME

"Nuts" can be played with small slips of paper and pencil, or the answers given verbally, a nut being handed to the first one who gives the correct answer. Here are the questions and answers:

1. What nut grows nearest the sea? (Beechnut.)
2. What nut is the lowest? (Groundnut.)
3. What nut is the color of a pretty girl's eyes? (Hazelnut.)
4. What nut is good for naughty boys? (Hickory.)
5. What nut is like an oft-told tale? (Chestnut.)
6. What nut is like a naughty boy when sister has a beau? (Pecan.)
8. What nut lives in a pen? (Pignut.)
9. What nut is like a goat? (Butternut.)
10. What nut might be made of stone? (Walnut.)

POTATO-PEELING CONTEST

A potato-peeling contest is sure to be jolly. Provide a sufficient number of medium-size potatoes and paring knives and all begin at a given signal. For a second contest allow the guests to carve a face or figure from the peeled potato.

THE BLARNEY STONE

The blarney stone adds one more novelty and is considerable fun. A stone should be treated to a generous bath of whitewash and be placed in the center of a large table. A round one is best. Tell how the fairies have placed a spell upon it and great good fortune will attend any one who succeeds in kissing it, after having been blindfolded and turned around three times.

A "THRILLING" PASTIME

Seat all guests upon the floor around a muslin sheet, in an absolutely dark room. While someone tells a blood-curdling story, illustrative objects are passed around underneath the sheet from hand to hand. A kid glove, stuffed with bran and soaked in ice water for an hour is as clammy a hand as one could wish. Peeled white grapes, icy cold, make "loose eyes"; cold boiled macaroni and spaghetti will represent muscles; a toy mouse and spider, a hot baked potato and a prickly burr, add interest; while a plaster of Paris skull could be added for the climax.

THE LITTLE DUTCH BAND

A game in which all join is called "The Little Dutch Band." The players sit or stand around the room in a circle. The leader assigns to each some imaginary musical instrument—horn, fife, drum, trombone, violin, harp, flute, banjo, etc. Some well-known but lively air is given out, and the band begins to play, each player imitating as nearly as possible the instrument he has been assigned.

All continues well until the leader suddenly drops his instrument and begins playing on that of another member of the band. At this the player whose instrument has been borrowed must change his attitude to imitate the instrument the leader dropped. This continues, the leader taking up the imaginary instruments of the various players, and they at the same time adopting the leader's instrument, the one he started with, not the one he has just dropped.

INDOOR SHOT PUT

Have five 1-pint [480-ml] fruit jars and fifty beans. Five persons play at a time. A jar is placed in front of each contestant, who receives ten beans. At a given word each contestant drops his beans, one at a time, into the jar from a height level with his chin. The one having the most beans in a jar wins.

PINS AND POTATO RACE

Arrange three large potatoes on a table at one end of the room, and request each member of the party to carry them to the opposite side of the apartment, using an ordinary pin to lift them with. Set a time limit for this feat in every case, and let all who accomplish it in the number of minutes allotted draw for the prize.

GRATITUDE GAME

Each person is given a paper and pencil and told to arrange a list of numbers down the side of the paper, numbering one to fifteen. The players are then to make out a list of things they are thankful for. Only humorous answers will be considered; all others will be ruled out later when the lists are read. The papers are then given in without names attached, and judges are appointed to go over the lists and decide the winning person. The papers are returned, each person taking one at random, so as to relieve him of the embarrassment of reading his own. After the fun is over a small prize may be awarded for the best list.

MUSICAL CHAIRS

The old games never lose in popularity, so you will find "Musical Chairs" very good for use for the party you wish to give your class. Arrange chairs in a long row down the middle of the room, placing them so that one faces one way, the next the other, and so on down the line. There should be one chair less than the number of players. Form a line, start the music (a bright march on the talking machine is just the thing), and when all are marching merrily around, stop the music. All scramble for seats, and the one who is left over stands aside, out of the game. Another chair is removed; the music starts up again and then stops suddenly. Again a player is left out, until it gets down to two players and one chair, the one who finally gets the chair wins the game.

LITERARY SALAD

A "Literary Salad" is also a great party game. For this prepare salad leaves by folding and twisting pieces of green tissue paper until they look like lettuce leaves; then paste slips of paper containing familiar quotations on these. The participants of this salad are requested to guess the name of the author of their quotation.

TREES OF THE WOODS

A list of words that suggest the names of trees may be prepared as follows: (1) twins; (2) to languish; (3) groom; (4) a tool, etc. The answers are: (1) pear; (2) pine; (3) spruce; (4) plane. These names are to be guessed. If there are musicians present, such old songs as "Woodman, Spare That Tree" and "A Rare Old Plant Is the Ivy Green" may be sung. For prizes, a dainty apron cut in the shape of a maple leaf and edged with lace would be pretty for the girls, while a shaving-paper case, also leaf-shaped, would do for the boys, and a book on trees for anybody.

GAME OF ANIMALS

Sides are chosen, whose members stand closely grouped by their respective leaders, X and Y. X calls out an animal whose name begins with *A* and counts ten. If Y can respond with another before X has finished counting ten, he does so, and begins counting, and X has to name an animal in *A*. This is repeated until no more names in *A* are forthcoming, when another letter is taken. If Y cannot give a name before the ten counts have expired, X chooses one of Y's followers, and vice versa.

When one side confesses its inability to name any more animals, its opponents are entitled to choose as many members of that side as they can give new names, beginning with the given letter. The only duty of the other players is to suggest new names for their respective captains. Before beginning this exercise, it would be well to appoint a secretary to make a list of the different names given, arranged alphabetically.

SUM CONTEST

For this, half the participants, standing still, hold a slate and pencil and write out an easy sum on the slate. The other half of the players, at the word *go*, run toward the slates, do the sum, and return to their places. The person doing the sum correctly in the shortest time is the winner.

GAME OF SCULPTURE

Here is a suggestion for a party small enough so that all can gather around a large table. Seat the guests. Place in front of each a card having on it an animal's name, five toothpicks, and a small lump of putty, from which each one must form the animal whose name is on the card, using the toothpicks for legs. Give them 5 minutes to do the work.

TOSSING THE SMILE

The main object of the game is to keep your face straight during the times when you are not "it." Ask all players to form a circle, either seated or standing. One person is chosen to stand in the center. Suddenly he smiles a broad smile at some one person in the circle, who smiles back, and the two exchange places. None of the other persons in the circle must allow their facial expressions to slip a mite, or a penalty will be exacted. A rapid exchange from the outside circle to the center necessitates alertness and interest on the part of the players who never know when they will be called upon to stand in the center of the circle and "toss the smile."

The following penalties for smiling out of turn may be used: Make the person go around and smile at everyone present, individually; have the person smile three times at himself for 30 seconds without stopping. Other penalties on this order may be used.

GAME OF WIRELESS TELEPHONE

On rainy days, or at other times when playing outdoors is impossible, the time can be made to pass pleasantly by playing "wireless telephone." The players are seated in a circle, preferably on the floor, and each one is supplied with a "wireless" outfit, which is simply a newspaper rolled to form a hollow tube about 18 inches [46 cm] long.

Everything being ready, the starter begins the game by whispering a message through his "wireless" to the neighbor on the left. The message must not be repeated if the neighbor fails to understand it clearly; but it is to be passed on to the next player just as it was received. When the last player has received the message he tells out loud what he was told, and then the player to his right also tells aloud the message as he heard it, and so on back to the starter. This is done to find out why the message did not reach its destination in its original form, and many ridiculous changes will be brought to light. The leader may say, "It is raining hard today," but by the time the last player is reached it may be changed to "Dick is straining hard and hay," with many other funny changes along the line.

"POOR PUSSY" AND "ROOSTER"

The one who is "it" gets down on the floor in front of one of the players and meows, spits, and cries like a cat. The person for whom this performance goes on must say, "Poor pussy, poor pussy," without smiling. Three "meows" are allowed, and if the person has not smiled, "poor pussy" goes on to the next one. Who ever smiles is "it." "Rooster" is played in much the same way, except that the players crow, flap wings, scratch the ground, etc. There is no response to the "rooster," the player must only keep from smiling.

YANKEE DOODLE KITCHEN

Divide the company into two equal groups; half the group will then take part in a "Yankee Doodle Kitchen," going through in pantomime the motions of washing clothes, scrubbing floors, churning, washing dishes, etc. The piano or orchestra plays "Yankee Doodle," at first very slowly, gradually increasing in speed, and the workers increase their movements accordingly until they are going at breakneck speed. This, while it sounds very simple, is lots of fun and must be seen to be appreciated. The second half of the group acts as audience and guesses the occupations. They then change places and the second group pantomimes the occupations of the early settlers—felling trees, picking up stones, plowing, sowing, harvesting, building, etc.

GAME OF MISSING ADJECTIVES

Someone writes up a story beforehand; anything will do, but the account of some former gathering, a picnic, drive, or walk affords the most pleasure. All the adjectives are left out. When the game is played, someone goes around the room, getting an adjective from each person, in turn, putting them into the blank spaces left in the story. When all the spaces are filled, the story is read, and it always proves a success.

A RING GAME

All join hands and form a ring, with the person who is "it" outside. They then move quickly around singing a tune as they do so. The person who is "it" taps someone on the shoulder. When the ring at once stands still, the person who was tapped runs one way, and the person who was "it" runs the other; when they meet, they bow to each other three times, then both try to reach the vacant place in the ring. The one who fails to bow three times, or fails to reach the vacant space in time, is "it."

COTTON CONTEST

Put dabs of raw cotton on the floor and let the players take them up on a teaspoon. Another version is to blindfold the players, throw beans or corn on the floor, and let the players gather them up in teaspoons. The one who gathers the most gets a small prize.

FISH TALK

Each player must talk like a fish. Of course we all know that the finny tribe has no speech, but the person who is "it" calls on each player in turn to make a noise like some certain fish, naming them in turn. The responses are ludicrous, but any one who fails to respond must be "it."

NEW FORM OF STAGECOACH

Give to each person the name of a city, town, state, or country, then let the one who is "it" call out quickly two of the names, such as "Boston" and "Galveston"; the person so named must quickly change places, while the person who is "it" must try to slip into a chair while the other two are changing. Whoever is left is "it." When the command comes, "All change cars," there is a grand scramble; all change places and the one who is left is "it."

MOVE THE PENNY

This is an indoor game that will always find great favor with a company of young people. The whole amusement is afforded by two balls about the size of billiard balls and a penny. It is necessary to mark out on the tablecloth, with chalk or pencil, a circle about 3 inches [7.5 cm] in diameter, and a straight line about 2 feet [61 cm] from the circle.

Put one ball in the center of the circle and on its top balance a penny. The trick is to bowl from the line with the remaining ball and try to knock the penny out of the ring. Simple as it may seem, it takes a great deal of practice, for nine times out of ten the penny will drop within the circle. The best way to accomplish this is to bowl very slowly, and, by knocking the ball very lightly, the penny will roll out on the top of the other ball.

ROMANCE

Provide each guest with a pencil and pad. The leader reads the topics aloud, and after each topic the paper is folded over so that no one sees what her neighbor has written; it is then passed to the right, so that each person writes in every booklet. The topics are as follows:

1. Name of the book.
2. Heroine's name.
3. Describe her fully.
4. Hero's name.
5. Describe him fully.
6. Where did they meet? Describe the place.
7. The heroine's impression of the hero.
8. What did the hero think of her?
9. The villain's name, describe him fully.
10. What terrible thing did he do?
11. What effect did his villainy have upon the heroine?
12. What did the hero do about it?
13. What did the heroine do next?
14. How were their difficulties solved at last?
15. What became of the hero?
16. What became of the heroine?
17. What became of the villain?
18. Moral.

At the end, each guest reads aloud the book that she has in her hand.

GAME OF HIDDEN CITIES

1. What city has few people? (Scarcity.)
2. What city is full of hypocrites? (Duplicity.)
3. What city has many chauffeurs? (Velocity.)
4. What city has greedy people? (Voracity.)
5. What city is for reporters? (Audacity.)
6. What city is for authors? (Publicity.)
7. What city is for wise people? (Sagacity.)
8. What city has great crowds? (Multiplicity.)
10. What city has odd people? (Eccentricity.)
11. What city has unhappy people? (Infelicity.)
12. What city is full of office-seekers? (Pertinacity.)
13. What city is for telegraph operators? (Electricity.)
14. What city is for the nations? (Reciprocity.)
15. What city is full of truthful people? (Veracity.)
16. What city is for talkative people? (Loquacity.)

OUTDOOR GAMES

OUTDOOR FAN TAG

Some member of the company is made "it" and receives a palm-leaf fan. He begins operations at once by trying to tag the others, which is done by touching them lightly with the fan. Anyone who is tagged, instead of becoming "it," as in ordinary tag, takes the hand of the person who tagged him or her and together they try to tag the others. The second person also receives a fan, which he passes to any one tagged by him and constantly playing at his left hand. This goes on as long as anyone remains free. The two persons at respective ends of the line carry palm-leaf fans and either end can tag. The fun of the long unwieldy line moving awkwardly around, endeavoring to tag a lively player who is determined not to be caught without a chase, is easily imagined. There is no prize to this game.

DAY AND NIGHT

The players are divided into two sides, standing about 6 feet [1.8 m] apart. One side is called "Day"; the other side "Night." A player in the center between the two sides tosses a card—black on one side, white on the other. If the black side comes up, Nights run and Days chase them. Days tag as many Nights as possible. Beyond a line 60 feet [18 m] away, the Nights are safe. If the white side of the card comes up, the Days run. The number of players tagged by the Days makes the Days' score, and the same rule applies to the Nights. Repeat this as often as desired, and add the scores on each side; the side having the highest total wins the game. The player in the center is scorekeeper.

"ARE YOU THERE?"

"Are You There?" is popular on shipboard, but can be played on land as well. Prepare two clubs of straw, winding them round and round with cord so that they will be firm. Blindfold two people, put a straw club in each one's right hand, then let them lie flat on the ground, clasping each other's left hands. One cries, "Are you there?" The answer comes "Yes!" But the one who answers immediately dodges the blow that follows, for the questioner tries to strike his contestant on the head with the straw club, and tries to locate his head by his voice. Each has a turn, and the one whose head is struck loses.

FOLLOW THE LEADER

In a company of bright boys and girls the old-fashioned frolic known as "Follow the Leader" is most enjoyable. One person is chosen as leader. Whatever he or she does is imitated by the rest of the players in a body. The leader may resort to tongue twisters in which he has already acquired some skill, or he may elect to lead the crowd in a merry dance, bounding over small obstacles, running around chairs, crawling under bushes, and any other mirthful stunt suggested by the occasion.

SCHOOL GAMES

SHADOW PANTOMIMES

Shadow pantomimes are very funny and easy to manage. The first thing to do is to select a poem suitable to illustrate. "Lord Ullin's Daughter" is a good one, and there are some excellent things in the *Bab Ballads* by Gilbert. A large curtain of white cotton cloth is stretched across the stage, with a light so arranged that everyone who passes between it and the curtain throws a shadow upon the latter. Someone who reads well is selected to read the poem slowly, and during the reading it is acted in pantomime.

AN ILLUSTRATED MAGAZINE

A large frame of wood, made to represent a picture frame and covered with gilt paper, is erected on the stage. There is a curtain that can be parted in the middle hung on the inner side, and a screen covered with dark cloth, just back of it, for the background. The magazine contains a poem, a story, and the advertisements. A good reader announces the name and the date of the magazine and the title of the poem; he then begins to read the poem, and this is illustrated with tableaux or pantomime, inside the picture frame. The story is read and illustrated in the same way; then follow the advertisements, which need no announcement. Soap, hair tonic, baby foods, chocolate, etc., anything that can be made into a "picture," can be advertised.

SPELLING GAME

This spelling game is entertaining. Form the company in a part circle; let someone begin with a letter, an *i* for instance, having in his mind independent. The next person must add a letter; but if he inadvertently adds a *t*, or even an *n*, he is out of the game, as either one of them makes a word. With greater precaution the next one probably adds an *m*, with *immediate* in his mind; but his next neighbor, thinking of improbable, quickly adds a *p* and is out; and so on.

A MOTHER GOOSE PARTY

This is pleasing for a home gathering, or for a church or school party. Let each child come dressed as a Mother Goose character. Use cheesecloth for costumes. It is cheap and comes in colors. Even crepe paper may be utilized by clever fingers. There are so many possible characters that, if the party is to have a guessing characters' contest with it, there is no need of duplicating if the children get together and decide on which character each will represent. Little Boy Blue needs only a blue suit and a horn; Simple Simon, any kind of a suit and a fishing pole. Let the guests guess who the children represent, and when the announcement is made have the Queen of Hearts present the one naming most of the characters correctly with a tart, given on a pretty tray, as a reward of merit.

PROGRESSIVE INITIALS

A number of tables to suit the number of guests are prepared, and are labeled in order, Fruits, Flowers, Noted Figures in American History, Clues of America. In the center of each table place twenty-four assorted letters, facedown. These may be pasted on small squares of cardboard, if desired. After guests are seated at each table, give to each a second card on which are written the subjects of the different tables. When all are ready, the bell at the first table rings and the game proceeds. One person turns a letter. If the first table is for fruit, and the letter turned is *A*, the person who turns it says "Apple," and keeps the letter. Then in quick succession a letter is turned by each person in rotation, until all the letters are exhausted.

The object of the game is to be the first to think of a fruit, flower, noted person, or city. When the letters are all exhausted at the first table, the bell rings and the game stops. The two persons who have gained the most letters during the game progress to the second table, and those at the next table who have the least take their places.

Each person keeps a record on his card of the letters he gets, and at the end of the evening a prize is given to the one who has held the most letters.

MIND READING

The so-called mind-reading experiments always promote fun. With the aid of an accomplice, one can mystify a roomful of people by offering to guess any number from 1 to 10, which they may choose, while you are out of the room. When you re-enter the room, you go to one after another of the players, placing the hands on each side of his or her head, pretending to get the number in that way. When you reach your accomplice, he compresses his jaw the number of times decided upon by the company. For instance, should they decide upon the number 6, your accomplice compresses his jaw six times so you feel it, and so "guess" the number.

Another mind-reading feat is to tell the players that you will guess any musical instrument they may pretend to play. Your accomplice brings the players in one by one, and they go through the motions of playing some instrument, which you will guess, although the player stands behind you and you are blindfolded. You will ask the player to perform a little faster, and when he is going as fast as he can, you can say, "It's coming, a little faster, please. There, I have it. You are playing the fool!" This is all good-natured fun and always raises a laugh.

GAME OF ADVERTISEMENTS

Pin around the room on the wall, well-known advertisements. Cut from them all printing and number each one. Let each one write down the numbers on a card, and opposite the number his guess. Whenever possible, pin up several advertisements of the same kind, numbered alike. This enables everyone to see the advertisement without crowding. To the one guessing the largest number give a small prize.

CHARADES

Charades are always fun, though familiar to almost everyone.
Choose sides. Let each side select a word that can be acted out.
If the other side guesses the word when it is acted, it becomes
their turn. Here are some simple characters:

> *Mistake*: A "miss" steps forward and "takes" something
> handed to her.
> *Bandage*: Let all come in pretending to play an instrument
> "band," and go out doubled over to indicate "age."
> *Mansion*: Let a "man" stand at one side while the others in
> passing turn aside and "shun" him.
> *Furlong*: All examine first a "fur" and then a "long" string.

To end the evening pleasantly, bring in a board on which is
pasted a large rabbit without ears. Blindfold some of the persons
present, and in turn hand each one an ear to pin on; turn the person about and give him the other ear. No two ears match, and the
results are very laughable.

For refreshments serve coffee and cake, and perhaps fruit.
Charge a small sum.

PROGRESSIVE CONVERSATIONS

Cards should be prepared with a list of subjects, such as "Should
the weather be tabooed as a topic of conversation?" "Do you
prefer the mountains or the seashore, and why?" "Which is the
greater educator, reading or travel?" "Is music without words
capable of expressing emotions?" The gentlemen's cards should
be larger than those of the ladies, and each pair, large and small,
should bear the same number. They should be distributed hap-
hazard, and the lady and gentleman having the same number are

matched for the first topic of conversation. For a fixed time, say 5 minutes, the first topic is to be discussed. Then at a signal, the gentlemen move on one place, leaving the ladies seated as they were. Thus each topic is discussed until the list is completed. At the close the gentlemen decide by vote who was the best talker among the ladies, and the ladies do the same for the gentlemen. If there is some simple prize, like a small dish of bonbons, it adds to the fun to have the victorious pair share it together.

THE POPULAR MAGAZINE

The contents of a magazine are represented by tableaux. The best arrangement is to have first the frontispiece, some pretty tableau or single figure; then a poem, with human illustrations—the text of course to be read by some good reader; then a story, followed by an illustrated article; and finally the advertisements, which are the most effective of all. The name of the advertised article is to be guessed from the tableau. This entertainment gives unlimited opportunity for ingenuity, skill in making up, etc.

LEFT-HANDED CONTEST

Have the right hand of every guest bandaged before entering the room, and throughout the evening insist on everything being done with the left hand. Another amusing game is to demand something in the way of entertainment from everyone present. A good way to manage this is to make a list of old-fashioned or well-known songs. Have four copies of each song ready, write the names of the songs on slips of paper, and let each guest draw one on entering the room. When ready, each group of four must sing the song that they drew. As the groups are sure to be mixed, the result is very entertaining.

HOME PARTIES

INEXPENSIVE ENTERTAINING

There was a time, and it was a good time, when one was satisfied to serve lemonade and cake, perhaps ice cream and, to be very fine, coffee, nuts, and raisins. We enjoyed it because it was good, and the desire-for-something-different microbe had not entered our souls. But now that it has entered, we seek and demand variety, and not to attain it is to confess oneself behind the times.

PUT YOUR WITS TO WORK

So, when you have a party, you must set your wits to work; and, while your refreshments and decorations may be very inexpensive, you must contrive, in some way, to have the latter suggest the objects of the entertainment or the time of year. If you prefer not to use linen, suitably decorated tablecloths and napkins made of crepe paper may be had for a small sum. You can also get, for a few pennies, paper cases to hold candles, nuts, etc., and favors suitable for any holiday and almost any purpose. Pretty china is not at all expensive nowadays, and all the glass necessary for the most elaborate entertaining may be had in the pressed ware, a pretty pattern known as "Colonial" being very inexpensive. A pretty plant, a few ferns, or some wild flowers will do for the bit of green that is now thought essential; while ferns laid flat upon the table are prettier than any embroidered centerpiece ever used.

WHAT TO SERVE

As for the food, if something substantial is desired, try broiled or creamed oysters, creamed lobster, or scalloped crabmeat, if you can get seafood; creamed chicken, if you live inland. With any of these serve thin slices of brown bread and butter, olives, and sweet gherkins. Or you can serve stuffed eggs with mayonnaise dressing and cold sliced chicken or turkey or a vegetable salad with thin slices of cold ham or tongue. You could also serve sandwiches with various fillings, such as cream cheese mixed with chopped nuts; sweet peppers, chopped fine, mixed with butter, lettuce, and mayonnaise; thin slices of cucumber dipped in French dressing; minced ham, tongue, chicken, or hard-boiled eggs; preserved ginger cut in very thin slices; chopped dates and nuts; chopped peanuts and butter; and so on indefinitely. The bread for sandwiches must be cut thin and is best when a day old. They must be neatly made to be appetizing. With the sandwiches may be served a salad. Chicken, lobster, shrimp, or crab salad may be used, or a simpler one made of celery and nuts, or celery and chopped apples, or a fruit salad. Olives and salted nuts may be served with any of the things mentioned. Omitting the sandwiches, you may serve crackers and cheese with the salad.

SWEETS AND DESSERTS

A sweet dish of some kind usually winds up the menu. This may be that old standby, ice cream, or orange jelly with whipped cream, or little individual chocolate puddings with hot chocolate sauce. Still another delicious sweet is made by adding marshmallows, cut in quarters, and halved white grapes, to sweetened whipped cream, served very cold. Small cakes are served with these sweet dishes. Rich cookies, cut in small shapes, or a good layer cake baked in small tins and iced will do.

Homemade candy is always enjoyed, and may take the place of the sweet dish, if desired. A tried-and-true recipe for fudge is as follows: Three squares of grated chocolate, 3 cups [600 g] sugar, 1 cup [240 ml] milk, and butter the size of a walnut. Boil all together until a drop of it placed in cold water will form a ball between the fingers, then add 1 teaspoon vanilla extract. Beat all until creamy, turn out quickly into a buttered pan, and cut into squares before it becomes quite hard. Another good recipe is for Mexican Panocha: Boil together 1 tablespoon butter, 4 cups [800 g] brown sugar, 1 teaspoon salt, and 1 cup [240 ml] milk. Cook this until it drops hard in cold water. Then pour in 2 tablespoons vanilla and 2 cups [240 g] chopped nuts, either pecans or walnuts, and stir constantly until well mixed. Pour on a buttered plate and cut into squares.

If the weather is cold, wind up with coffee or cocoa; in warm weather serve lemonade or grape juice.

GUESSING BY FEEL

Place on a table about twenty or twenty-five bags, each numbered and containing some well-known article, such as a comb,
bar of soap, apple, potato, spoon. Each person may handle
each bag once, and must handle them all in a given time. Write
down each player's name and the number of bags guessed correctly. The one guessing most may receive an inexpensive prize.

SMELLING CONTEST

Into each of a dozen or more colored bottles is poured a small
amount of some liquid that can be detected by the odor—
turpentine, vinegar, witch hazel, camphor, peppermint, coffee,
cologne, bay rum, wintergreen, etc. The bottles are then ranged
in order on a table, and each member of the company provided
with pencil and paper, passes in turn by the bottles, giving a sniff
to each, and then from memory makes a list of the contents of
the bottles. For the most perfect list a prize is awarded. This
game abounds in fun. Sniffs must be brief; and as perfect solem-
nity must be preserved for accurate results, the spectacle of the
sniffing procession is deliciously absurd.

THIMBLE CLUB ENTERTAINING

It will be very pleasant to start your club with a social afternoon,
and you could play the following games either indoors or on the
porch: For a Button Contest, arrange and number your tables.
On the one marked No. 1 there are, for each person, fifteen
large buttons with thread and needles. On the other tables, there
are bowls filled with buttons. The person at the head table who
sews on her fifteen buttons first rings a bell and progresses with
the one who has sewed on the next highest number, first pulling
off the buttons so as to be ready for the newcomers. Make a knot

in the thread, sew once into each hold, then fasten enough to hold the button on. Break the thread each time. Every person reaching the head table sews on the fifteen buttons as the first did, the remaining persons beginning over again, and keeping the score. Those at the other tables sew on just as many buttons as possible, while the ones at the head table are doing the requisite number.

After fifteen progressions, the cards are collected, the score counted, and prizes are awarded the must successful contestants. After this, distribute cards with the following misspelled words written upon them: 1. *dhetar* (thread); 2. *yreemgba* (emery bag); 3. *stbutno* (buttons); 4. *nisp* (pins); 5. *eednsle* (needles); 6. *eatpeursema* (tape measure); 7. *koosh* (hooks); 8. *seey* (eyes); 9. *xaebesw* (beeswax); 10. *sfeipnatsy* (safety pins); 11. *tgrisn* (string); 12. *halck* (chalk); 13. *hrtesdiakl* (silk thread); 14. *ardib* (braid); 15. *scssrois* (scissors); 16. *elbmith* (thimble); 17. *debsa* (beads); 18. *nragdintoocnt* (darning cotton); 19. *knodbi* (bodkin); 20. *ranettp* (pattern); 21. *dugro* (gourd); 22. *inkdleeettngin* (knitting needle); 23. *blueck* (buckle); 24. *grsa* (rags). Tie a pencil to each card and allow the guests to write the correct word opposite each misspelled one. Simple prizes may consist of articles needed in the sewing room.

THIMBLE PARTIES

Thimble parties are popular this year, when groups of women come together to sew for the needy. Invite your friends to luncheon, which may be a simple affair. Serve clam broth, creamed chicken, rolls, sweet pickles, salted peanuts, chocolate cornstarch with custard sauce, little sponge cakes, and coffee.

In the center of the table have a work basket filled with ferns, and, with the dessert, bring on a pretty glass dish of strawberries (emery bags, of course) to be given as favors. The following contest may be arranged on cards, and guessed during the luncheon:

1. What does the farmer do to his sheep—shears.
2. To pick one's way—thread.
3. What is thrown away—waist.
4. A sign of servitude—yoke.
5. A berry—thimble.
6. A blow—cuff.
7. A company of musicians—band.
8. Deep-sea animal and part of his body—whalebone.
9. An exclamation—a-hem!
10. A kind of music—piping.
11. Necessary to hang a picture, and part of the human body—hook and eye.
12. A piece of furniture and a weight—cotton.
13. Money and a derogatory adjective—cashmere.
14. A grassy yard—lawn.
15. Preposition and a fisherman's term—overcast.
16. What the cook does to the turkey—baste.
17. A part of an eatable animal—mutton leg.
18. Part of a door—panels.
19. A negative—knot.
20. A prejudice—bias.

Tie a small pencil to each card, using narrow white tape for the purpose, and present the workbasket centerpiece to the person making the largest number of correct guesses.

A MARRIED FOLKS PARTY

In a gathering of married people, each man might write a description of his wife's wedding dress, while the wives might describe any difficult experiences they had met with, such as ironing "John's" shirts, making "pies like mother used to make," preparing for unexpected company when the larder was empty, etc. A pretty centerpiece is in the nature of a Jack Horner pie, but may be called "Cupid's Pound Cake" instead. Suitable favors are wrapped in tissue paper and tied with red ribbon or string. These strings are pulled through a hole in the center of the (paper) crust, and one is carried to each place. The sides of the "cake" are decorated with paper hearts or cupids, and at a given signal each guest pulls a string and finds a favor at the end of it.

To find one's dancing partner for the evening, have a number of pretty cards in a fancy basket; write the name of a lady on each. The men each take out a card and must take for a partner the lady whose name appears thereon.

HAT MAKING

Provide each lady with a paper napkin and ten pins, with which to make a hat in 5 minutes. Let the ladies put the hats on and march around, while a jury decides upon the prettiest hat. The one making it receives a prize—a paper of pins or an equally insignificant article.

A CAT PARTY

A cat party will be funny and novel. Cut out of magazines, advertisements, etc., all the pictures of cats, large or small, that you can find, and draw, paint, or trace a cat on your invitations. Cut pieces of cardboard about 6 by 14 inches [15 by 35.5 cm] in size; at the top of each one paste a cat picture and mark it in large numbers.

Underneath in large letters, write or print a sentence representing a word that has for its first syllable *cat*. For example:

1. A midnight cry.
2. A list of books.
3. An ancient engine.
4. A raft with a sail.
5. A waterfall.
6. A mountain range.
7. One who furnishes food.
8. An accident.
9. One of the finny tribe.
10. One of the furry tribe.
11. To ask questions.
12. A sauce.
13. A common herb.
14. An instrument of punishment.
15. A dupe.
16. An insect.
17. Part of an instrument.
18. A cold.
19. A church.
20. A jewel.
21. One of the feathered tribe.
22. A religion.
23. A common weed.
24. A burial place.
25. Grazing animals.

This list can be added to indefinitely, the dictionary furnishing several pages of words beginning with *cat*. Pin or stand the cards in conspicuous places about the room, and furnish each guest with a pad and pencil. The answers are numbered and the person guessing the largest number receives a small prize.

It will be quite easy to obtain suitable prizes for a small sum; if not, a penwiper or needle book made of black cloth, cut in the shape of a cat, will do.

The game of "Stuffed Cat" is next in order. Two leaders are chosen; they form the company into two lines of equal numbers. One line is called "face," the other "back." One of the leaders throws a stuffed (calico) cat into the air. If it falls on its back, the "face" line laughs loudly, the other line keeps still; if it falls on its face, the other line laughs, etc. If any one laughs on the wrong side, he must forfeit a player to the other side, and the game goes on until all the players are on one side.

Pieces of black tissue paper, about 6 by 7 inches [15 by 17 cm], are then passed around, and each player is expected to make a cat out of his piece. He must tear it with the fingers, as no scissors are allowed. When done, they are mounted on sheets of white paper and the guests vote for the best and the poorest. If there is room, the old-fashioned game of "Pussy Wants a Corner" can then be played, and for those who would rather keep quiet, provide string for playing "Cat's Cradle." For refreshments, serve cookies cut with a cat-shaped cutter and Catawba grape juice.

PROGRESSIVE NOVELTY PARTY

A progressive novelty party is easily arranged and enjoyed by young and old. Arrange as many tables as will be required, seating four persons at each table. At table No. 1 have some sort of a card game, such as Flinch, Old Maid, Snap, etc. For the other tables you can have peanut stabbing, using hatpins to remove the peanuts from a bowl; needle threading; the new jackstraws, using magnets and steel nails, the latter of various sizes, and the game played just as the original jackstraws was played.

A miniature game of ring toss can be prepared for another table, using a spool and skewer (the latter driven into a smooth pine board) for the stake, and the rubber rings that come for fruit jars for the rings. Table croquet can also be managed with a homemade outfit; a smooth pine board with the spool and skewer stakes, crossed wire nails for hoops, mallets made of corks and skewers, and mothballs will be sufficient for a good game. Number your tables; allow 5-minute intervals of play; and when the bell rings, those who have won move up to the next table; and so on indefinitely until the end. Simple prizes may be given to the winners.

After this let all join in the game of "Flying Angel," seating all the players save the leader in a semicircle. Give the leader a white handkerchief and ask him to pass it to the player at one end of the circle. The game is to throw this handkerchief from one player to another without it being caught by the leader, who may stand wherever he chooses or run about from player to player. If the leader touches any player while the handkerchief is in his possession, that player must take his place as leader.

SPECIAL OCCASIONS

FEBRUARY ENTERTAINING

The birthdays of Washington and Lincoln, together with St. Valentine's Day, afford many suggestions for the hostess. For Lincoln's birthday, the guests might come dressed as members of the Grand Army of the Republic and their families. War songs and plantation melodies could be sung, and someone could read the Gettysburg Address aloud. Decorate the table with a cheap plaster bust of Lincoln, with streamers of red, white, and blue crepe paper, extending to the corners. Serve army beans (baked), salt horse (corned beef), pickles, hard tack (biscuits), cake, and coffee.

ST. VALENTINE'S DAY ENTERTAINING

Decide on your date and send out invitations from 10 days to 2 weeks in advance. If the fourteenth is the chosen day, a roll of red wallpaper, cut into hearts of various sizes, will help along in your decorations, and an occasional arrow of gilt paper may be thrust through the hearts. There are other suitable decorations—favors, napkins, etc., to be had from paper manufacturers for a small sum, and you can use the heart device through the entire evening. Start with a heart hunt, using paper hearts, candy hearts, etc., with a prize for the one finding the largest number, and a mitten for the booby prize. The larger hearts can be cut in two pieces, and these are matched to form partners. Pass tablets and pencils, and ask each man to write a proposal of marriage to his ideal, while each girl writes an acceptance to her ideal. These should be well mixed up and then drawn from a hat, and a proposal and acceptance read together. For refreshments serve creamed chicken in heart-shaped paper cases, or heart-shaped sandwiches, little cakes cut heart-shaped, or a large layer cake with heart-shaped candles to decorate the top, and a ring, a thimble, and a piece of money hidden in the cake. The one who gets the ring will be married first, the thimble goes to the one who will remain single, while the money brings wealth.

A VALENTINE ENGAGEMENT LUNCHEON

Last year a girl who was about to announce her engagement did so on Valentine's Day. She entertained her girlfriends at luncheon, the table being suitably decorated with a center-piece that told the story. Ferns, asparagus vine, and red tulle were prettily massed around a black velvet cat emerging from a bag of red silk. Around its neck was a red ribbon, and hanging from this was a white card containing the names of the newly betrothed pair. "The cat was out of the bag" at last. Heart-shaped place cards, strings of red paper hearts draped around the sides of the table and from the chandelier to the corners of the table, and sandwiches, cakes, and candy in heart shapes all carried out the idea.

ST. VALENTINE GAMES

Follow the heart idea in your games. Old Maid may be played with a set of cards each of which is pasted on a red cardboard heart. Play "Hearts Up" with a tiny heart, just as you play "Up Jenkins." Have the men guests each write a description of the ideal lady of his heart, and the girls in turn write descriptions of their ideal mates.

A VALENTINE SOAP-BUBBLE GAME

Suspend a large sheet across one corner of the room and on it paste three large red paper hearts, numbering them 1, 2, 3. Above each one write a small verse. The first:

Blow your bubble right on here,
And you'll be married before another year.

Above the second write:

To be engaged this very week,
Number two is the one to take.

Above the third write:

A sad, an awful fate awaits the one who seeks me,
For he or she will ever a spinster or a bachelor be.

On a small table nearby have a large bowl filled with soapsuds and also clay pipes decorated with hearts. Small paper fans should be given to each player, who first blows the bubbles off his pipe, then tries to fan them on the heart where he wishes them to go. Most will try to avoid heart number three.

YOUR HEART ON YOUR SLEEVE

One person out of the assembled company retires from the room. Those remaining behind choose a state of mind, such as "Joy." The person outside is called back. When he counts 1, 2, 3, those taking part in the game strike an attitude representing "Joy." The person called in then tries to guess what they are representing. The first person who laughs while the attitude is being assumed is sent out after the player guesses the word to be "it" next time. Each guesser has three chances; if all three guesses are wrong, he goes out again.

Some suggestions for words that can be acted out are *anger, indifference, jealousy, pity, curiosity, stupidity, pride, expectancy, disgust, fear, self-consciousness,* and *dignity*.

To match partners, provide a basket of cardboard hearts and, on arrival, require each boy to punch one of them with a key in his possession. Distribute the punched hearts among the girls. Find partners by matching keys and the keyholes.

VALENTINE PROPOSALS

As the guests assemble, give each gentleman a slip of paper bearing the name of a woman, and the ladies the name of some man noted in fiction as a lover. Thus the one who has Romeo hunts for the lady who has Juliet on her paper. When all know who their partners are, the ladies must evade every attempt on the part of the gentlemen to propose to them during the evening.

A prize is given to the gentleman who has succeeded in proposing, and to the girl who has eluded all efforts of her partner by her wit and sagacity.

FOR WASHINGTON'S BIRTHDAY

For the twenty-second of February, ask guests to be dressed in old-time costumes and have the rooms lighted with candles and sperm-oil lamps. A collection of relics may be arranged for old songs and games indulged in, and a real colonial supper served. Serve tea and old-time cookies, and if it is a money-making affair, sell tea and china on commission.

FOR ST. PATRICK'S DAY

Cut from green paper a number of pieces approximately representing the map of Ireland. There are as many of these as guests, and to each a little pencil is attached with ribbon. Each player is given one, which he or she is called upon to fill out with the names and positions of the various large cities, rivers, mountains, etc. A book bound in green makes a suitable prize.

CHALKING THE PIG'S EYE

If possible, draw a pig on the floor; if the floor is not suitable, draw it upon a blackboard, or, using charcoal, upon a sheet. Then blindfold each player, turn him around three times, and tell him to mark the pig's eye with a cross. A variation on this is to have the players draw the tail, which should be omitted in the original drawing. Still another fun-making scheme is to pass sheets of paper and have each person, while blindfolded, or with closed eyes, draw a pig. Suitable prizes for such contests are the little brown earthenware piggy banks, to be had for a few pennies each.

APRIL FIRST ENTERTAINING

Some clever games for entertaining on All Fool's Day may include the "Foolish Walk," for which pile sofa pillows, books, plants, and anything in the way of obstruction on the floor; then tell a certain person to mark each article carefully in mind, blindfold him, and tell him to walk across the room. In the meantime, after the victim is blindfolded, the objects have been noiselessly removed, leaving the floor clear. It is amusing in the extreme to see the blind one making his way, and when the bandage is removed the astonishment is great.

This may be followed by a guessing contest. Prepare several amusing collections of objects meant to symbolize particular dinner courses beforehand. Provide cards and pencils for each guest, with numbers for each course of a dinner menu. Bring your prepared "courses" in one at a time, and after 2 minutes remove. The contestants write down the name of each course as they guess it, and a prize is given to the one making the largest number of correct guesses.

The following "dishes" are suggested: oysters, short pointed ends of blue crayons (blue points); soup, small brown cardboard turtles, in a soup bowl of water; crackers, tiny firecrackers; meat;

a toy lamb in a small pan; poultry, a map of Turkey with the name erased; cake, the ends of sulfur matches (devil's food); nuts, the iron nuts used in bolts and machinery. Decorate with vegetables instead of flowers, and among the refreshments have a dish of chocolates, which are nothing but cotton-batting, dipped in melted chocolate.

HALLOWEEN

Old games, old customs, and old tricks and charms are appropriate for Halloween entertaining, and the gathering can take the character of a "Hard Times Social," the guests wearing their oldest clothes (a prize may be given for the very oldest), and if the company can be accommodated in a large kitchen, so much the better.

Use wrapping paper and the cheapest envelopes procurable for the invitations, and arrange the table with white oilcloth or a colored cotton cover. For a centerpiece fill a toy wooden washtub with rich red apples. Around this arrange candles stuck in potatoes or carrots for holders. Tin pie plates, or the wooden picnic kind, may be used, with paper napkins and tin cups. Refreshments may be simple—sandwiches, salad, salted or plain peanuts, gingerbread, doughnuts, cookies, molasses candy and coffee; or, if a hot supper is desired, you can choose between baked beans or scalloped oysters, rolls, pickles, individual pumpkin pies, coffee, and nuts.

Cornhusks will hold salad. The nuts should be brought to the table in a great wooden bowl. This should be placed in the center of the table and the guests will be asked to help themselves. However, some will be found tacked to the bottom of the bowl, two of the guests will find their nuts fastened together by means of a tiny wire or thread, and all kinds of confusion will result. When the nuts are opened they will be found but empty shells, the kernels having been removed to make way for small bits of paper on which is printed the fortune of the finder.

HALLOWEEN FORTUNES

Here are some Halloween fortunes, short and optimistic, that
may be used:

For you will come bright, happy days.

You will never marry unless you are suited.

Profit will attend your ventures.

Your companion in life will be ever true.

You have genius, but must develop it.

You will not become wealthy, but you will never want.

Early in life you will know honors.

You will wed the one you love.

Continue unafraid of work—it is not afraid of you.

*Of course there are sorrows in your life, but they are balanced
by joys.*

Never spend money foolishly—you cannot earn it foolishly.

You will travel extensively.

Your wealth will come from the earth.

A companion worthy of you will enter your life.

HALLOWEEN SUGGESTIONS

Have the guests enter the house through a cellar door. Light the cellar by means of pumpkin lanterns, and have a ghost meet them and silently motion them toward the cellar stairs. Instead of ducking for apples, which wets the hair, have two pieces of stick, sharpened to a point at each end, and these nailed together to form an *X*. On the four points are stuck, respectively, an apple, a potato, a piece of soap, and a piece of candle. The *X* has a piece of string caught into the nail in the middle, and is suspended from the chandelier and set spinning. Then you stand around and try to bite at the apple as the cross spins. If you bite the apple, it is a sign of a rich and early marriage; if the potato, you marry a farmer; the piece of soap means you marry a poor man; and the candle is to light you as you wait for your husband to come home.

The old tricks never lose interest, but for the sake of novelty may be changed. A piece of candle in the tub of water may take the place of the apples for which young people enjoy "bobbing."

COSTUMES FOR HALLOWEEN

The list of possible costumes is endless. Sheets and pillowcases, with a white or skull mask, are easily arranged for a ghost costume for either boys or girls. A witch requires a dark woolen skirt, a black cape, and a wig of coarse hair hanging in strings from beneath a black pointed hat. She carries a broom, of course, and a black cat made of paper may be perched on her shoulder.

For a rag-doll costume, take two pieces of muslin each about 14 inches [35.5 cm] long and 11 inches [28 cm] wide, and round the corners. Sew up on three sides; paint nose, eyebrows, and mouth on it; and cut out places for the eyes; slip this over the head. Wear white cotton gloves, and wear stockings over your shoes and a cotton dress made with a long skirt. Practice walking in a loose-jointed, floppy way, to carry out the illusion.

A baby costume is easily fashioned by wearing a rather full nightdress over long white petticoats. A mask representing a baby face, a bib, a white cap, and rattle complete the costume.

As for the boys, a slender lad dressed as a girl is always a success. Uncle Sam, cowboys, and pirates are always popular and are easily copied from pictures.

Witch

Rag Doll

CHRISTMAS HOLIDAY ENTERTAINING

Delightful entertainments may be given during the holiday season, when the young folks are home from school or college. A gathering of friends and neighbors, old and young, may be entertaining and instructive and yet inexpensive. The decorations consist of the Christmas greens, bells, etc., and the tree as the center of attraction. There are many pretty cards at this season that may be used for invitations and place cards. The shops are full of toys and novelties that may be used as favors and prizes, and the refreshments may be very simple, with Christmas cakes playing an important part in your menu.

A TWELVE-MONTH SOCIAL

All young people, and some not so young, love to dress up, and an interesting affair can be made out of a Twelve-Month Social. Ask your friends to come dressed or wearing some device to represent the months of the year, and offer prizes for the best ideas. January may come as Father Time; February offers a wide choice with its famous birthdays and the feast of St. Valentine; March offers the hare and "Paddy"; April brings Easter; May is the blossom month and is also sacred to our dead heroes; June brings roses; July is, of course, patriotic for Americans; August offers the summer girl; with September comes Labor Day and the "whining schoolboy, with his satchel, and shining morning face." October might be represented by Columbus; while November brings us Thanksgiving Day, instituted by the Pilgrims; and the year winds up very properly with Christmas and Santa Claus.

CHARADES

A Christmas or New Year's dinner could be enacted in charade form, as guests take turns acting out the courses as best they can for the audience. Each person gets a copy of the menu, with only the courses written down, and must guess at what constitutes each course from the charades. The best guesses receive a prize. The menu, with some suggestions for how best to act out these foods, is as follows:

Soup, noodle (new-dull); roast, turkey (Turk-key); gravy, giblet (Jib-let); vegetables, potato (pot-eight-o); cauliflower, (call-i-flower); succotash (suck-at-ash); jelly, currant (cur-rant); dessert, plum pudding (plumb); beverage, coffee (cough-fee).

The old game of "Consequences" may be varied for the occasion, called "Resolutions," and played accordingly. If the party is held on New Year's Eve, it may wind up with the birth of the New Year, finding all standing in a circle with joined hands. As the clock strikes twelve, the company sings, "Should Auld Acquaintance be Forgot," and then with a handshake and a greeting for every one, the party breaks up.

GAME OF RESOLUTIONS FOR NEW YEAR'S

Provide guests with papers and pencils. Begin by having ten letters of the alphabet read to the company. These are to be copied down and the guests must choose a new year's resolution of ten words, each beginning with one of the letters used in the order in which they have been given out. These impromptu resolutions when read will cause much amusement.

THINGS *for* IMPATIENT CHILDREN TO DO

MAKING A MEGAPHONE

Tom and Elizabeth are setting a gate at the far side of the 20-acre [8-hectare] pasture. You want Tom to bring in the monkey wrench, so you swing your hat and bawl at him for fully five minutes. Finally they both drop their tools and come in empty-handed.

It wasn't done that way at the Fort Myer training camp. Our company, split up into platoons, might be drilling all over the huge parade ground; but the West Point instructor, with his megaphone, could control the most distant unit as easily as the nearest one. After I came home—discharged for a small physical defect—I made several useful megaphones.

The lid of a large pasteboard box furnished the material. I worked out the pattern. In one corner of the box lid, I drove a pin with a string tied to it; by looping the other end of the string around a pencil, I drew the two curves at the distance I have figured.

Cut out the piece that is shaded and roll it up; lap the edges about ½ inch [12 mm] and glue them so. Then hold it with your thumb and first finger crooked around the small end, which is pressed against your lips to form a sort of mouthpiece. A sentence spoken in a firm, distinct tone will carry a wonderful distance.

A couple of thick coats of shellac will waterproof the megaphone. A large megaphone, naturally, will carry farther. Still, the one shown will answer most folks' needs.

MAKING A HATSTAND

Even laymen can easily make this useful piece of furniture. Any kind of wood will do, but use oak if possible. It requires one upright 66 inches [167 cm] long and 2 inches [5 cm] square; one base 20 inches [50 cm] long and 2½ inches [6 cm] square; two bases each 8¾ inches [22 cm] long and 2½ inches [6 cm] square; and four triangle braces 4 inches [10 cm] across the base and 8 inches [20 cm] high.

On the 20-inch [50-cm] base, directly in the center, cut a 4-inch [10-cm] square, mortise 1 inch [2.5 cm] deep, into which fit the upright. Nail it securely. To each side, nail or glue one of the side bases. Then nail one of the four triangle braces to each base and against the upright. Screw on four clothes hooks at the top. The tree is then ready for a coat of varnish or paint.

Note: By using small-headed finishing nails, driven below the surface with a nail set, and then puttying the holes, the wood will not split nor will the nail holes show.

MAKING A SNOW HOUSE

We are sorry that all of our children haven't the opportunity to build a snow house. It's lots of fun, and if the house is properly made, it will last for a long time. First of all shovel some snow in a big pile, stamp-  ing it down until it is compact and hard. Make the pile at least 8 feet [2.5 m] high. Then at one side start to hollow it out. The entrance can be small, so that one has to creep in, or it can be made large. If the snow has been carefully packed, the roof will not cave in. The snow brought out can be piled at the sides of the entrance as a windbreak. If the nights are very cold, a little water can be sprinkled over the outside of the house. This will freeze and the house will last longer. We'd like to see a good photograph of your snow house and have a full description of it.

MAKING A CARDBOARD LANTERN

Secure some good tough cardboard. Make a base 12 inches [30.5 cm] square. Score it 1 inch [2.5 cm] all around, cut at each corner, and turn up edges. Make and fasten securely in the center a tin holder for the candle. If you can't make this, an ordinary tin candlestick can be used. The top is made the same as the bottom except that in the center a hole 5 inches [12 cm] in diameter is cut. The sides are 12 inches [30.5] square; score and turn in 1 inch [2.5 cm]; cut out the center. Cover openings on all four sides with yellow or red tissue paper. Any design desired can be painted or pasted on the tissue paper. Paint the outer side of the cardboard black. Glue the lantern together or fasten with wire. Attach a wire to each of the four upper corners, join, and hang the lantern from a pole. See that the candle is held securely, and watch out that the top or sides do not catch fire.

MAKING A SHAMPOO BOARD

A useful article is a shampoo board, made from a board 8 by 10 inches [20 by 25 cm]. Stand the board on end and, about 1½ inch [4 cm] from each upper corner, begin to cut down a curve that must be shaped to fit the neck. Cover the edges of the curve with rubber from an old inner tube.

A brace to make the board stand up may be sawed from a block 1½ inches [4 cm] thick and 6 inches [15 cm] square, making the cut diagonally across it.

The use of this board when washing long hair avoids the tangling sure to follow when the hair is thrown forward over the head in the usual way. Place the board in position in the sink or a box of proper height. The user leans back, places her neck in the padded curve, letting her hair fall down in its natural position in the sink. The shampoo is quickly given by another person without tangling a single hair—a point not to be overlooked.

A TREE HOUSE

There probably is at least one tree on every farm that can be used as a base for a tree house. A good strong tree, with branches spreading in such a way that boards can be easily laid from branch to branch, should be used. If properly built, there is no danger, and much enjoyment can thus be had. Build it strong; take no chances. Remember, safety first.

A HANDY TWINE HOLDER

"Do you know where I can find a piece of string?" How often this question is asked! Well, here we have an idea that when properly worked out will give us an ornament, a good-luck omen, and a good twine holder.

First sketch the front and back parts of a cat. Trace these on a piece of ⅜-inch [1 cm] poplar wood and cut along the outline with a jigsaw. Sand both pieces until they are smooth on both sides and on the edges.

With a hard pencil draw eyes and a nose on the front half of the cat. Paint the back half black, and all of the front half excepting the spaces for the eyes and nose, as indicated. When the black paint dries, color the spaces for the eyes green with a black pupil. Paint the nose pink. When both pieces are thoroughly dry, varnish them with clear shellac.

After the front and back halves of the cat are finished, measure a piece of wire ⅛ inch [4 mm] in diameter and file or grind both ends to points. This acts as a support for the spool that revolves on it. Force these pointed ends into the wood by carefully hammering the wood onto the points. This should be done before the cat is fastened onto the base.

Put the spool in place and wrap on the twine (any color). Make a wooden base for the completed twine holder and fasten by nails or screws that pass up through the feet.

Varnish the entire wooden object, and your article is complete. If the body is short, the effect will be better than if too long. Paint white whiskers on each side of the nose.

A TREE FOR CLIMBING

Climbing is a natural instinct, and a tree for climbing should be in every playground. A straight tree trunk about 30 feet [9 m] high, with the bark removed and made smooth but not necessarily even, should be used. Plant it securely in the ground and protect the top by a platform sufficiently wide not to allow its edge to be grasped by the climber, who must not get on the platform. Only one child to climb at a time. Take turns and play fair, and be sure an adult is on hand to supervise.

THE BALANCING TREE

This can be easily made for your community playground. Cut down a large and perfectly straight tree, free it of the bark and limbs, and round it off; it should be 50 feet [15 m] or more long. At the thicker end the tree may be 2 feet [61 cm] or more in diameter. It tapers to an end of 4 to 6 inches [10 to 15 cm] in diameter, which is to be free to sway. It is supported by two securely nailed wooden feet (which are anchored in the ground), one at the extreme thick end, the other one sufficiently far from the thinner end to allow the thin end free play to bend and sway. The tree is so supported that at its thicker end its upper edge would be 36 to 42 inches [91 to 105 cm] from the ground. This tree, as its name implies, gives a chance for balancing exercises on a broad and steady and also on a more narrow surface, which sways and bends. It is not so dangerous as tightrope walking and just as much fun. Always see that the ground on both sides of the tree is free from stones, and never push anyone off. Take your turn and play fair.

HORIZONTAL BAR

Here is a horizontal bar suggested by the Playground Association, and it can be easily made. Two 2×6 [5×15 cm] planks 7 feet [2 m] long and a 1¼-inch-by-6-foot [3 cm by 2 m] steel pipe are required.

Sink the planks securely in the ground. Bore holes in them to admit the ends of the pipe; the pipe should be left free to be raised or lowered to a higher or lower level by moving it to higher or lower holes. The pipe should be kept from turning by putting a bolt or pin through both plank and pipe at right angles to the other holes in the plank. Your blacksmith will help you out. Instead of pipe, the crossbar can be made out of fine second-growth hickory, thoroughly seasoned and nicely finished. Some prefer steel bars; others wood. In either case use good material to prevent accidents. Under the bar have plenty of soft dirt, free from stone.

A "SAVE YOUR SHOT" TARGET

This target allows shot to be used over and over, dozens of times. With it the air rifle can be used in the house without any danger of injuring the walls, furniture, etc. Upon a wooden support 48 inches [1.2 m] high, place a box. This may be made any size desired. I built one with the front, or target, 24 inches [61 cm] square. The slanting back is 36 inches [91 cm], base 12 inches [30.5 cm], and the tapered conductor, which runs the shot into the can, has its sides 16 inches [40.5 cm] long.

The back and sides of the conductor are lined with felt. The target is a bull's-eye drawn on a cardboard 24 inches [61 cm] square, and the cardboard is tacked upon the front of the target box. A BB shot passes through the cardboard, hits the slanting back, bounds against the conductor side, and drops into the can behind the target. The slanting back, with its felt lining, kills the force of the shot without battering it, and so the same shot can well be used dozens of times.

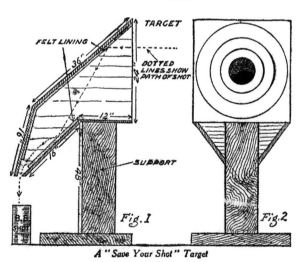

A "Save Your Shot" Target

TUB TILTING

Tub tilting on land is equally as exciting and requires just as much skill as tilting on the water. Secure two barrels, about flour-barrel size, and two poles. Each pole should be 8 to 10 feet [2.5 to 3 m] long, of the lightest possible wood, with a big soft pad on the end. These are spears for attacking. The barrels are set level, exactly at poles' length apart, center to center. Each contestant takes his place on a barrel, and he must try to put the other fellow off. The umpire stands alongside, near the middle. For safety's sake it is a good idea to have someone stand behind each player to act as a catcher in case of accident.

It is counted a foul to push the other player below the knees, to use the spear as a club, to push the barrel, or to take hold of your opponent's spear with your hand. A foul gives the round to the other contestant. A round is up when one player goes off his barrel. If one drops his spear and can recover it without getting or falling off, it is all right.

A battle usually lasts for about seven or eight rounds. The best players gain their points by wriggling their bodies and keeping in continual motion. There is a lot of fun and excitement in keeping your balance.

GAME OF SMUGGLERS

The group must be divided as equally as possible and a captain chosen for each side. One side takes the part of the "smugglers," the other side the part of the "laws." The game can last as long as desired. One day and one night is a good length, but it can be made longer or shorter. Decide on the time before the game begins. A line must be fixed and two marks made on it about ¼ mile [400 m] apart. The smugglers camp on one side of the line and the law on the other. Camps should be hidden as much as possible. The smugglers must land their goods across the line between the marks made. The goods of the smugglers can be so many sacks (as decided upon) filled with sand or other weighty material. The players should watch each other's camp and report to their respective officers. The side of the law must (unknown to the smugglers) watch the line between the camps all the time.

The players can report in person, by a companion, or by lights; never by calling, as that would give the position away. A very good line over which to land the goods is a small creek. One-quarter mile [400 m] in length is about all that can be watched successfully. At night sentries should be changed every 2 hours, oftener if possible. Two must watch the line at one time. If the smugglers land their goods undiscovered, they win. It is a good game and you must use your wits.

FOOT IN THE RING

Here is a good game for cold weather; any number can play, but it is best to have squads of eight. For each squad draw on the ground a circle about 2 feet [61 cm] in diameter. Player No. 1 comes forward, places one foot in the ring, bending the knee and having the weight of the body over this foot. He then folds his arms and waits the attack of Player No. 2, who, also having her arms folded, hops forward. No. 2 hops around No. 1, who keeps changing his front to where No. 2 is, until she finds a chance to attack No. 1, and while hopping, push him out of the circle. If she succeeds, she wins, and takes the circle, No. 3 coming forward to attack her, and so on. If, however, during the contest No. 2 gets both feet on the floor, she loses, and No. 3 then comes forward to attack No. 1. The player in the ring, so long as his foot is in the circle, may cause the attacker to fall by evading or dodging him. The arms must always remain folded and the pushing must be done with the shoulders, and never with the raised arms. An exciting contest is had by putting two attackers against the one in the ring.

THE SNAKE CHASE

It is said that St. Patrick drove the snakes out of Ireland, and here is a game that will help celebrate the event. Any number can play. One player is chosen "keeper of the snake." The rest are "chasers." The snake is made of a log 10 inches [25 cm] long and at least 6 inches [15 cm] in diameter. Fasten three horseshoes securely on each end, and around the log drive spikes. Fasten a rope so that the log can be dragged by the keeper.

This snake leaves a trail which is easily followed on ground, but which is much harder to follow on grass or through woods. The keeper of the snake starts out 10 minutes ahead of the chasers. The snake must never be lifted from the ground, but the keeper can retrace his steps and do everything possible to confuse the chasers. Three places should be chosen and known to all as the snake's home, and the object is for the keeper to get his snake to one of these homes before he is captured by the chasers, who must follow the trail, and who cannot leave it to capture the snake.

GIANT STRIDE

The apparatus can easily be made and should be in every community playground. It is the next best thing to flying. A strong old wagon wheel, a pole 18 feet [5.5 m] long and 5 inches [12 cm] in diameter at the smallest end, and 60 feet [18 m] of 1-inch [2.5-cm] hemp rope are needed. In almost every district someone will provide the wheel, and the pole can be cut in the woods. If the wheel has a wooden axle hub, remove the axle from the skein, which is the metal sleeve surrounding the axle spindle to protect it from wear. Shape the top of the pole to fit into the axle skein, then fasten the skein securely in place.

If the wheel has a metal axle, get, if possible, a blacksmith to help you. Have the axle cut off about 1 foot [30.5 cm] from the hub and sharpen it to a point. Into the middle of the small end of the pole, bore a 2-inch [5-cm] hole 6 inches [15 cm] deep and drive the axle into it. Then have an iron collar shrunk around the pole to prevent it from splitting. An all-metal wheel and axle is better than a wooden one.

Cut the rope into four lengths of 15 feet [4.5 m] each, and with copper wire or by splicing attach the four ropes to the hub. Knot each rope every 2 feet [61 cm] from the bottom for a distance of 6 to 8 feet [2 to 2.5 m]. Set the pole securely 4 feet [1.2 m] in the ground in concrete. Cover the hub with a tin shield to protect it from the weather. Have the ground around the base of the stride free from stones.

How to use the giant stride: Catch hold of the rope, start to run around the pole, and the momentum will soon take you off your feet.

POM-POM-PULLAWAY

This is an excellent recess game, as it takes just about 15 minutes to play it:

A child stands out in the middle of the long school yard, or some other open place where two lines can be drawn about 50 yards [45 m] apart. All the other children range themselves on one of the lines, or against the school yard fence. Then the child who is "standing" yells, "Pom-pom-pullaway!" All the children on the line run for the other line, past the child in the middle, who tackles any runner that she thinks she can catch and hold. As soon as she has brought one of the children to a stop, that child is compelled to join her in "standing," and these two then call the others who have got by to the other line. "Pullaway!"—the rush is resumed, and probably two more children are caught; and so on until all are caught; and there is always a terrific struggle at the end, because the biggest, swiftest, and strongest are hardest to stop; and by that time these difficult ones have the whole school tackling them. Pom-pom-pullaway is good football practice.

PART V

OUT
in the
WORKSHOP

A STORAGE CELLAR *for* ROOTS *and* VEGETABLES

Concrete is one of the most satisfactory materials for storage cellars, not only because of its permanence, but because its use makes it possible to maintain definite temperatures. When the frame, roof, and walls are made of wood, brush, or wire, the storage cellar has to be rebuilt every few years because of the changes in the moisture content of the soil coming in contact with the wood or wire. The moisture causes the wood to decay and the wire to rust. The best wood cellar is only temporary.

A root or vegetable storage cellar of concrete is built to last. Like a concrete silo, a concrete storage cellar of proper size should pay for itself in a year. Marketing can be controlled in strict accordance with supply and demand and the most favorable conditions, and waste by rot entirely prevented, if crops are stored carefully and at the right time.

The storage cellar described here and shown in the accompanying diagram is suitable for either fruit or root crops. Its size can be varied to suit orchards or farms of different sizes, simply by adding to or taking from the length, leaving the other dimensions unchanged. The plans show a cellar 12 feet [3.5 m] wide, 14 feet [4 m] long, and 9 to 10 feet [2.75 to 3 m] deep inside. Floors, roof, and walls are of concrete, with tar joints where the walls join the floor. The cellar is moisture-proof and secure against rats and mice.

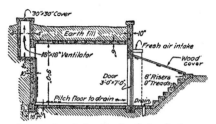

A section through the cellar, lengthwise, showing ventilator and fresh air intake, together with dimensions.

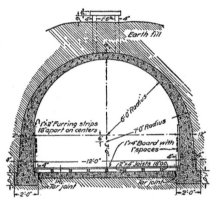

A section through the cellar, crosswise, showing how the false floor and walls of boards are put in. Notice the tar joints to allow for expansion.

End walls are 10 inches [25 cm] thick. Side walls are 16 inches [40.5 cm] at the base and taper to 6 inches [15 cm] at the crown. The floor is 5 to 6 inches [12 to 15 cm] thick. The materials required for a cellar of the size shown in the plans are forty-five barrels of cement, 14½ cubic yards [11 cubic m] of sand, and 22½ cubic yards [17 cubic m] of pebbles. The concrete for the walls is mixed in the proportion of one part cement, two and a half parts sand, and four parts pebbles or stone. The floor and arch of the roof are made of a 1:2:3 mixture. The floor is sloped toward the entryway where the drain is located.

This design has been prepared with special reference to ventilation. During cool evenings manhole and cold air intake covers are removed and the cold air permitted to pass down the intakes, circulating through the passage between the concrete floor and the false floor of the bins. The false floor is made of 2×4 [5x10 cm] joists, covered with 1×4 [2.5×10 cm] boards nailed 1 inch [2.5 cm] apart. Openings in the floor allow the air to pass up through the stored contents, thus cooling them.

Outside walls are built so that cool air can circulate up along them. The warm air passes out through the manholes. In the course of one night, all of the air in the storage cellar is in this way changed many times, thus thoroughly cooling the cellar before morning.

The cellar holds about 600 bushels of potatoes. On the average farm, where the whole cellar is not needed for vegetables, it can be divided by a partition and part of it used for other purposes. During the summer months it can be used for a milk cellar or a storm cave.

INDIVIDUAL HOG HOUSE

A man from the Simpson place drove past his neighbor's farm with five sows in the wagon one morning in early April. His neighbor was fixing the fence at the foot of the hill, just where the man stopped to rest his team.

"Got them from the Joneses' farm," came from the driver in answer to his neighbor's question. "I don't know where I'm going to put them, though."

"Well, what you ought to do is to make some movable sheds, one for each sow, and pull them out into the lot so the sows can give birth undisturbed. I use them almost entirely; they're cheap. Drive on up the hill, tie your team, and come and see them."

"Now, here's one in use," he continued as they walked about the hog lot. "This sow farrowed out here three days ago, all alone. Saved every piglet."

"Looks as if those would be easy to make," observed the new man.

"Easiest thing in the world. You can make one in an hour or two. See, you take two pieces of 4×4 [10×10 cm] 8 feet [2.4 m] long for runners and build a floor on them, using 2-inch [5 cm] planks for floor. Then use a 2×4 [5×10 cm] 9 feet [2.7 m] long for the ridge, and 2×4s [5×10 cm] 8 feet [2.4 m] long for the slanting studs, making an A-shaped frame. Put the roof boards up and down, and hinge them on one side for two doors that can be opened toward the sun on bright days. Between the upright pieces of 2×4 [5×10 cm] you can make a door at one end. I've made mine to slide up and down."

"What would five of those cost me?"

"That depends on what grade of lumber you use. I'll give you a list of the lumber for one of them, and you can take it to the lumberyard for prices."

After figuring on the roof of one of the houses for a few minutes, he produced this list, which he copied on the back of an envelope and handed the new neighbor:

Two 3-foot [91-cm] 4×4s [10×10 cm] for runners
Two 9-foot [2.7-m] 2×4s [5×10 cm], for ridge
Three 3-foot [91-cm] 2×4s [5×10 cm], for studs
Four 4-foot [1.2-m] 2×4s [5×10 cm], for end studs
One 2-foot [61-cm] 1×4 [2.5×10 cm], for above end door
One 10-foot [3-m] 2×4 [5×10 cm], for fenders
Nine 6-foot [1.8-m] 3x12s [7.5 by 30.5 cm], for floor
Eleven 16-foot [4.9-m] 1×10s [2.5×25 cm], for shiplap for roof
Four 12-foot [3.6-m] 1×10s [2.5×25 cm], for shiplap for ends
Four 3-foot [91-cm] 1×4s [2.5×10 cm], for door battens
Three pairs 6-inch [15-cm] strap hinges
One pair 4-inch [10-cm] strap hinges; nails

HANDY
PIG CATCHER

With this very handy pig catcher, one can scoop a pig weighing up to 30 pounds [13.6 kg]. If dropped in front of a running pig, he will go into it of his own accord and will not squeal. An old-fashioned hind-leg grab isn't necessary with this catcher. To make one, take a piece of ⅜-inch [1-cm] round iron cut to about 4 feet [1.2 m] or a little longer, heat, and bend it into a hoop, allowing about 4 inches [10 cm] of each end to project for insertion into a shovel handle with a heavy ferrule (a metal band that binds the tool to the handle and prevents the handle from splitting). The net is made of a heavy burlap sack, secured to the round iron frame with heavy twine in such a way that three-quarters of the sack shall hang down and form the net.

THE BARN
I'D LIKE *to* HAVE

I have a small barn on my own place, but I'm not going to show the plans. Why? Because the former owner built it, and he didn't know much about planning a barn; that was plain. So here's the barn I'd *like* to have—not the one I have, unfortunately.

The first story is exactly 22 by 44 feet [7 by 14 m]; the broad front should face south, to give the cows the greatest possible amount of warmth and shelter in winter. Each cow stall is 42 inches [1.2 m] wide, in the clear; the horse stalls are 54 inches [1.5 m], or possibly 1 inch [2.5 cm] or more for large draft horses. Nine feet [2.7 m], including the manger, is a good length.

The box stall about 10 feet [3 m] square is absolutely necessary for a sick horse or a cow with a calf. The shop is nearly the same size. Harness, saddles, etc., can be kept here. The wagon house, 16 by 22 feet [5 by 7 m], will hold two wagons, two carriages, and various other things very easily. Your windows in the wagon house admit plenty of light.

The wall studs are 2×4s [5×10 cm], set about 2 feet [61 cm] on centers, with 4×4 [10×10 cm] sills and 2×4 [5×10 cm] plates. The center posts are 6×6 [15×15 cm], carrying a girder of three 2×12 [5×30.5 cm] joists. The second-story joists are 2×10 [5×25.5 cm], set 2 feet [61 cm] on-center. In most cases they are 12 feet [3.5 m] long, but above the stalls half of them must be 14 feet [4 m] long, because there the girder isn't in the exact center of the stable. The rafters are all 2×4s [5×10 cm], 12 feet [3.5 m] long; the diagonal braces are 1×6 [2.5×15 cm], about 12 feet [3.5 m] long; while the little short pieces are 1×4 [2.5×10 cm].

I should advise using 6-inch [15-cm] lapped siding for walls and the gables; the roof may be shingled or covered with some sort of patent roofing. Ventilators may be put on the top, though a simpler scheme is to make slatted openings high up in either gable. The foundations had best be of concrete. Cork bricks are good for stall bottoms.

AN EMERGENCY AUTO REPAIR LIGHT

In traveling at night the motorist is often forced to change tires or repair a puncture when he has failed to take along a light. If a repair must be made that takes only a short time, a good emergency light may be had by heaping up a small mound of sand or dust, moistening it with gasoline, and setting it on fire. The mound should be made at a safe distance from the car. In order to avoid having a widespread fire, the mound should be depressed in the center, like the center of a volcano, with the gasoline poured in the depression. This will insure the best possible light under the circumstances. The motorist also owes it to property owners along the highways to wait until the light burns out, or to see that it is entirely extinguished before leaving it. Never try to replenish such a light with more gasoline, unless you are in a hurry to join the angels.

CARRY *an* EXTINGUISHER

A few weeks ago, Jim Caldwell met with an accident that was both lucky and unlucky. Jim's luck was the indisputable fact that he happened to be in town at the time instead of 4 or 5 miles [6.5 to 8 km] from nowhere. Jim had been tinkering with the carburetor of his auto, but its disposition kept getting worse and worse. It spit and it popped and it missed, and suddenly it went off like a cannon and flames shot up clear through the hood. Some gasoline had collected in the underpan.

A Handy Extinguisher Saved this Auto

Fires don't appeal to Jim and he didn't fancy sitting on the gasoline tank, so he jumped out and made tracks down the street. Somebody rang a fire alarm. Then Jim saw a garage man in greasy overalls with a fire extinguisher under his arm running toward the blazing auto. Jim went back to help.

Well, the hose and ladder wagon came, but the man had the fire out long before that. The paint on the hood was scorched and the chemical had made a lot of sediment on the motor, but Jim's auto had been saved from going up in smoke. Jim peeled a greenback from his roll and handed it to the hero of the occasion.

The auto was badly damaged. Before Jim drove home that night he went to the hardware store and bought a fire extinguisher that's been hanging on the dash ever since.

BOB'S ENDLESS CLOTHESLINE

"I wonder if Mother knows how tired I get lugging this old basket of clothes over the ground beneath this clothesline," grumbled Bob.

"I wonder if you know how tired your mother gets every week walking along under the line hanging up clothes and taking them down," replied Uncle John. "Why don't you fix up an endless line for her so she won't have to walk back and forth, and so you won't have to lug that basket around!"

"An endless line? What do you mean, Uncle?"

"Didn't you ever see an endless line? Well, we can easily rig up one. Got any old pulleys? Got any extra wire for clothesline? Got a couple of strong timbers for posts?"

"We've got about anything you can think of lying around the farm somewhere," replied Bob, interested at once.

Then they went to work. First, they hunted up a big pulley and fastened it to a beam just inside the laundry room window. Then they found a piece of wire that suited, ran it through the pulley, fastened the end loosely, and stretched the wire along the ground to find out where to set the post. When the post had been set firmly, they found another pulley and attached it to the top of the post. Then they loosened the ends of the wire, ran one end through this pulley and then, after drawing up the slack, carefully spliced the two ends together.

"Now get a couple of boards, a hammer and some nails," said Uncle John, "and come into the laundry room. I want to show you something."

In the laundry room they set up a strong platform about 2 feet [61 cm] high, running from beneath the clothes wringer to the window under the endless line. Uncle John picked up the clothes basket and placed it on the platform beneath the wringer.

"Now suppose that basket is full of wet clothes," he said. "No one needs to lift it. Just slide it along the board to the window. Now we're ready to hang up the clothes. Don't have to stoop over for every piece when the basket is on this platform. Where are the pins, boy? Get a small box and nail it to the wall, put the pins into it, and they'll always be within your mother's reach. Well, we'll proceed to hang clothes. As we pin them on the line, we'll move the line along. Better oil those pulleys a bit. When your mother sees how fine this works, she'll want another, because two short ones are better than one long one."

"That's a fine idea!" exclaimed Bob. "Mother won't have to go out in the cold. And—and I won't—"

"You won't have to lug that basket around anymore," Uncle John said. "Don't ever be ashamed to admit that your head has saved your legs and arms. It's not the hard work itself that counts; it's the results."

PICKETING *the* CALF

When no pasture can be provided for the calves, they can be picketed so as to have plenty of feed without too cumbersome a rope, by having two picket pins joined by a smooth wire, to which by means of a swivel is attached the calf's rope so that it will slip on the wire. When new grass is necessary, move the picket pins alternately. The length of the wire will determine the size of the grazing space.

COLLAR
for
SELF-MILKER

This is a simple device to be applied to self-milking cows (cows that suck their own teets). Since it is not patented, any of our readers are at liberty to use it. It is simply a necklace made from old broom strung on a strap and buckled around the neck. It should be fitted to the cow and the sticks made long enough to keep her from putting her head on her side and not long enough to chafe the shoulders or throat when the head is not turned.

PROTECTING *the* FARM HORSE'S HEAD

The straw hats used for horses' heads lack ventilation. Here is something to keep old Dobbin's head cool when working in the hot sun. It is simply a piece of paste board and a bit of wire, and can be made in two minutes. When bent into position, the ends of the wire slip into the long side loops in the bridle. This device allows the air to circulate under the pasteboard and still keeps off the sun.

SCRAPER *for* HORSES' LEGS

This device is better than a scraper whittled out of wood to clean horses' legs with. Have a broom made narrower and shorter than usual, sewing an extra seam in near the end of brush, and trim out a space in center of brush in shape of crescent, 2½ by 2 inches [6 by 5 cm]. This will do the work thoroughly.

SAFETY
PIG FENDER

"She is a good brood sow, but she always kills about half the pigs by lying on them." Haven't you often heard that expression from your neighbor? Next time you hear it, tell your neighbor how to prevent such trouble by fixing up a pig fender. Putting your brood sow in this device will separate her from her smaller piglets and prevent any unfortunate accidents. Tack a piece of 2×6 [5×15 cm] on the wall 6 inches [15 cm] from the floor by means of some 4-inch [10-cm] spikes. Then place three-cornered blocks above it every 3 to 4 feet [91 cm to 1.2 m] and put ¼-inch [6-mm] bolts through. The pigs will be safe every time.

WATER HEATER
and
FOOD COOKER

Take a barrel, an iron tube about 3 inches [7.5 cm] in diameter and perhaps 10 feet [3 m] long, and a wooden plug. Have thread cut in one end of pipe; screw it into right-size hole in barrel, 4 inches [10 cm] from the bottom. Drive the plug into the outer end of the pipe. Put water into the barrel. Build a fire under the pipe (dig a trench for the fire, if necessary). You'll have scalding water in about 30 minutes. When cooking feed, it is desirable to have a piece of wire fly netting fastened over the inside end of the tube.

HOMEMADE FEED CARRIER

Where there is a long row of cattle to be fed a grain ration, this device will prove labor-saving. Have the feed room at one end of the line of stalls, where the car can be filled. Then with a measuring scoop in one hand, the feeder can pass rapidly down the line, pushing the car before him. Any blacksmith can mount the little wheel in a frame with hook attached.

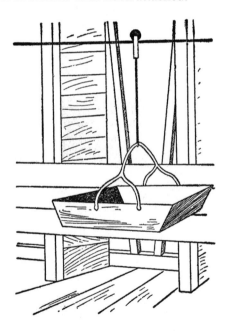

MILK PAIL HOLDER *for* CALVES

Bang! went the calf's head into the milk pail, and the milk splashed in every direction when the pail tumbled over. The man lost his temper and the calf missed a meal. And all the trouble might have been prevented had the man driven four pieces of 2×4 [5×10 cm] into the ground and set the pail inside them when feeding the calf. This plan saves milk, time, and temper, all of which farmers need to save.

KEEP ROOSTERS *from* FIGHTING

Take a stiff cord and make a loop long enough to hang from the bird's neck nearly to the ground. Fasten the end of the cord around his neck. With one end of the cord secured around his neck, the other end with the loop will dangle nearly to the ground, right around foot level for the rooster. When he makes a jump for his antagonist, his foot catches in the loop and over he goes, and he soon concludes he doesn't know anything about fighting.

KEEPING *the* CHICKS UNDER *the* TREE

No two things about a farm go together better than fruit trees and chickens. Make them acquainted early. Put the newly hatched brood in a coop under a tree and surround the whole with a circle of 2-foot [61-cm] poultry netting with 1-inch [2.5-cm] gaps in the mesh. It will stand alone in the form of a circle. The shade is good for the baby chicks. The chickens are good for the tree. As they grow, they will scratch the soil and thus cultivate and enrich it, besides destroying many insect enemies.

SIMPLE CLOSET

Small closets are indispensable about the barn, stable, and work-shop. Get a box at the grocer's in which fancy crackers come and screw it against the wall. It has a cover already hinged on. Put in one or more shelves, according to the use that is to be made of the closet.

GRAIN CHEST
from
PIANO BOX

At almost every village music store there are to be had empty boxes in which organs or upright pianos have been packed. These boxes can be utilized very nicely for grain chests. The narrow shelf gives accommodation for grain measures and feed boxes.

LANTERN HOOK

The farmer who knows what is good for him does not set the lantern down on the barn floor, where it may be accidentally upset. Here is the proper way to secure a lantern for both safety and convenience. At the back of the stalls, and elsewhere in the barn where needed, string an overhead wire at a height within easy reach. On each wire are one or more notched bits of wood. Then all that's necessary is to hook on the lantern and slide it along to just where it is wanted.

SMALL DOOR
in BIG ONE

Needless effort is constantly being used in opening and shutting big barn doors for the entrance and exit of a single individual. Cut a small door in one of the big ones and hinge it, putting a latch at the other edge. One can thus go in and out with small effort, and in winter when the big doors are drifted in, the small door, opening inward, gives ready access to the barn without shoveling away the drifts.

HANDY RACK *for* IMPLEMENTS

At one side of the shop have a rack for tools and garden implements. The garden fence is no place to hang a hoe when not in use. It's important to have a serviceable rack that holds multiple tools, including compartments for small tools, such as hammers, trowels, dibbles, or the like.

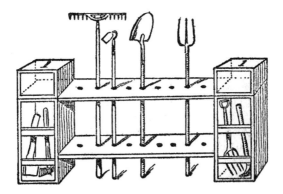

BURNING OUT STUMPS

Dragging out tree stumps by expensive machinery is slow and costly work, and in many cases may be avoided by drawing away the soil from the stumps so they will dry in the sun and wind, and then firing them in the following manner: With a 2-inch [5-cm] auger, bore through a large root near the ground on the side toward the prevailing wind; then with a squirt can moisten the inside of the hole with coal oil and get a fire started so the wind will draw through it. In this manner, if a draft can be maintained, the stump may be gradually consumed. Two people can fire and watch to keep the burning going, night and day, and as quickly clear an acre [4,000 sq m] as by the more forceful machinery or dynamite.

MICE PROTECTION FOR TREES

The writer has proved to his own satisfaction the real worth of old newspapers wrapped about the base of young fruit trees as winter protection against mice and rabbits and sunscald. An added protection can be had in the form of an old tin can wrapped about the base pressed down into the earth. A lot of old empty cans may be put into the fire and the solder melted off. The sides can then be pulled apart at the seam. If newspapers or cans are not at hand, use strips of lath placed upright close together and tied into position, or strips of wood veneer, or fine wire metal screen, or odd pieces of tin, sheet iron, etc.

A SIMPLER SCARECROW

Construct a wooden box, without top or bottom, and to the sides fasten bits of broken glass. Suspend this by a cord so it can swing freely. It will constantly keep flashing and dazzling in the sunlight, and will inspire much awe in the minds of crows and other predatory birds. Bright tin cuttings from the tin shop hung up in the field serve much the same purpose, but must be renewed as they rust at the cut edges.

NEWSPAPER
SHADE
for
PLANTS

When transplanting plants during warm weather, shade them with a little tent made instantly of a sheet of old newspaper. Fold the paper over the plant and secure the edges by dirt or stones, or by bits of lath with a stone in the middle. The paper turns the sun's rays, while the open ends permit a free circulation of air.

A RAKE *and* STONE BOX

Ground or small stones can be easily cleared with this simple device. An iron garden rake, and a box with one side removed, gives a "broom and dustpan" arrangement that makes the picking up of stones an altogether different affair from the old-fashioned finger and basket method. The box has handles to permit emptying it into the cart that is to haul the stones away.

IMPROVED SMOKEHOUSE

There are smokehouses of several types for curing but none so safe and satisfactory as the one in which no fire is ever put. It has a 6-inch [15-cm] tile running from a fire pit in the earth 3 to 8 feet [91 cm to 2.5 m] from the house and a trifle lower.

The smoke comes in at or near the bottom of the house and reaches the hams and bacon perfectly cool. Another advantage, the meat may be smoked without unlocking the smokehouse.

DOUBLE BARN BROOM

A single broom does not sweep wide enough a swath in the big barn floor, but two old brooms put together make quick work. Two old brooms with broken handles can thus be utilized, making a handle for the combination out of a broken shovel or fork whose handle is intact.

PRESERVING FENCE POSTS

The best preservative for this purpose is coal tar or creosote, boiled into the part of the post to be placed in the ground. Any tank or large can of sheet iron that will allow the liquid to be heated to the boiling point and that is deep enough to allow the posts to be covered with the liquid to a height of 30 inches [76 cm] will do.

Posts to be treated should be thoroughly seasoned, and treated for about 2 hours. If you can do no better, you can certainly put on some kerosene with a brush, letting the liquid soak into the wood as much as possible. Charring fence posts is not so effective.

HANGING BOARD *for* TOOLS

Tools are more easily found and more likely to be kept in place if they are hung up than when all are thrown together in a chest or drawer. Have a large board with a hole in the top to hang it up by. Hold up each tool against it and drive in stout nails in the place most convenient to hang it up. With hatchet or hammer this would mean two nails under the head with the handle allowed to drop between them. Then draw or paint upon the board the exact outline of each tool in its place. You will then know just where to hang it, and if one is missing you can ask with assurance, "Who has taken my hammer?" or "Where is my gimlet?"

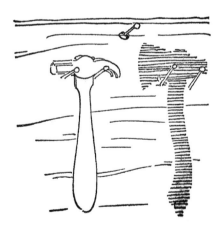

SILENCING *the* INVALID'S WATCH

The ticking of a watch or clock is sometimes objectionable in the bedroom of a sick or nervous person. It can be muffled in the following manner: Procure a large clear-glass jar with the neck big enough to admit a watch or small clock. Then by means of a rubber band suspend the watch from the cover midway in the jar. The time can readily be seen without removing the watch.

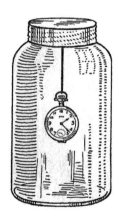

SHAVING MIRROR

No mirror is so good for shaving as one placed in the middle of a window, for the face gets all the light that the mirror reflects. Hence, the man of the house will welcome such a device. One pane of the bathroom window is removed and a pane cut from a broken mirror is put in its place. The original pane is put in behind it to protect the quicksilver, and both puttied in.

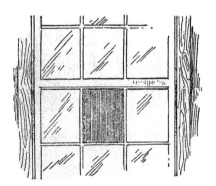

INDEX

ACKNOWLEDGMENTS

Thank you to the team at Chronicle Books, who brought this old book back to life in the most beautiful way. We are honored to work with such a thoughtful, distinctive publisher.

A special thank-you to Rachel Hiles, who immediately understood the vision for *How to Do Things* and has been its greatest champion ever since. The editors of *Farm Journal* would have been delighted to have found a kindred spirit in you.

Thank you to Brian Barth, whose foreword has added immeasurably to the enjoyment of this book.

Jay Venables and Chris Petronio deserve a very big thank-you for their tireless work restoring and editing the original manuscript. You both have a keen editorial eye, and this book would not be what it is today without your help.

Thank you to the past writers and editors of *Farm Journal,* who ultimately deserve all the credit for the book you hold in your hands.

And finally, thanks to all the hardworking farmers and country folk across this great land who know how to do things, and do it cheerfully. Long live the generalists!

AUTHORS

Brian Barth is the writer-at-large for *Modern Farmer* magazine and has contributed to publications including *Horticulture* magazine and the *Washington Post*, and sites such as the *New Yorker* online. Born in Georgia, he recently moved to Toronto, Canada, by way of a decade spent in the coastal hills of California. His weak spots include seed catalogs, baby goats, and heritage hog breeds. Farmers are his superheroes.

William Campbell is part outdoorsman, part writer. He resides on a hundred-acre farm in the Blue Ridge Mountains with his wife and an Australian shepherd named Tonto. He spends his days hunting, fishing, and reading old issues of *Farm Journal*, but, mostly, he just lives for a living.